Springer Theses

Recognizing Outstanding Ph.D. Research

For further volumes:
http://www.springer.com/series/8790

Aims and Scope

The series "Springer Theses" brings together a selection of the very best Ph.D. theses from around the world and across the physical sciences. Nominated and endorsed by two recognized specialists, each published volume has been selected for its scientific excellence and the high impact of its contents for the pertinent field of research. For greater accessibility to non-specialists, the published versions include an extended introduction, as well as a foreword by the student's supervisor explaining the special relevance of the work for the field. As a whole, the series will provide a valuable resource both for newcomers to the research fields described, and for other scientists seeking detailed background information on special questions. Finally, it provides an accredited documentation of the valuable contributions made by today's younger generation of scientists.

Theses are accepted into the series by invited nomination only and must fulfill all of the following criteria

- They must be written in good English.
- The topic of should fall within the confines of Chemistry, Physics and related interdisciplinary fields such as Materials, Nanoscience, Chemical Engineering, Complex Systems and Biophysics.
- The work reported in the thesis must represent a significant scientific advance.
- If the thesis includes previously published material, permission to reproduce this must be gained from the respective copyright holder.
- They must have been examined and passed during the 12 months prior to nomination.
- Each thesis should include a foreword by the supervisor outlining the significance of its content.
- The theses should have a clearly defined structure including an introduction accessible to scientists not expert in that particular field.

Amalio Fernandez-Pacheco

Studies of Nanoconstrictions, Nanowires and Fe$_3$O$_4$ Thin Films

Electrical Conduction and Magnetic Properties. Fabrication by Focused Electron/Ion Beam

Doctoral Thesis accepted by
University of Zaragoza, Spain

 Springer

Author
Dr. Amalio Fernandez-Pacheco
Cavendish Laboratory
University of Cambridge
JJ Thomson Avenue
CB3 0HE Cambridge
UK
e-mail: af457@cam.ac.uk

Supervisors
Prof. M. Ricardo Ibarra Garcia
Department of Condensed Matter Physics
University of Zaragoza
Pedro Cerbuna 12
50009 Zaragoza
Spain
e-mail: ibarra@unizar.es

Prof. José María De Teresa
Department of Condensed Matter Physics
University of Zaragoza
Pedro Cerbuna 12
50009 Zaragoza
Spain
e-mail: deteresa@unizar.es

ISSN 2190-5053   e-ISSN 2190-5061

ISBN 978-3-642-26696-6   ISBN 978-3-642-15801-8 (eBook)

DOI 10.1007/978-3-642-15801-8

Springer Heidelberg Dordrecht London New York

Cover design: eStudio Calamar, Berlin/Figueres

Printed on acid-free paper

Springer is part of Springer Science+Business Media (www.springer.com)

Supervisors' Foreword

The new era of nanosciences and nanotechnologies demands new materials and processes with large potential impact across several fields and guided by an interdisciplinary research. The great advances in preparation, characterization and patterning of materials allow today the observation of new phenomena and the generation of new devices. Major applications of these materials and devices can be already found in new branches of medicine for early diagnosis and selective drug delivery. For energy applications, new nanotechnological approaches for its cleaner and more efficient generation are being investigated. Highly-miniaturized data storage, processing and transmission is currently being applied to personal computers and mobile phones thanks to recent developments provided by nanotechnology. Nanomaterials are also finding many applications in pharmacy, cosmetics and construction.

Electron- and ion-based techniques are very important for the growth, visualization and nanoprototyping in nanosciences and are gaining relevance in the last years. In the present thesis work by A. Fernandez-Pacheco, the reader can find how the use of focused electron and ion beams allows the observation, etching and fabrication with nanometric control. In particular, a new technique based on electron and ion beam induced deposition (FEBID and FIBID respectively) is producing significant results and is expected to bring about new developments and phenomena. During his thesis work, the author has produced magnetic (Co-based) and superconducting (W-based) nanodeposits with potential applications in the fields of magnetic sensing and storage as well as in superconducting-based fluxonics. Another relevant result of the work is the etching of a thin film conductor down to the atomic scale using in-situ resistance measurements. One can conclude that the newly developed approaches to tailor magnetic, superconducting or metallic circuits with tuned conductance constitute new and exciting research directions with large potential for novel technological developments.

Zaragoza, November 2010

Prof. M Ricardo Ibarra Garcia

Prof. José María De Teresa

Contents

Acronyms

AF	Antiferromagnetic
AFM	Atomic Force Microscope
AHE	Anomalous Hall effect
AMR	Anisotropic Magnetoresistance
APB	Antiphase Boundary
APD	Antiphase Domain
BAMR	Ballistic Anisotropic Magnetoresistance
BE	Binding Energy
BMR	Ballistic Magnetoresistance
BSE	Backscattered electrons
CCR	Closed Cycle Refrigerator
CIP	Current In-plane configuration
CPP	Current Perpendicular-to-plane configuration
CPU	Microprocessor
CVD	Chemical Vapor Deposition
DER	Diffusion Enhanced Regime
DOS	Density of States
DW	Domain Wall
EBJ	Electrical Break Junction
EBL	Electron Beam Lithography
ECJ	Electrochemical Junction
EDX	Energy Dispersive X-Ray spectroscopy
ELR	Electron Limited Regime
ES-VRH	Efros-Shklovskii Variable Range Hopping
FCC	Face centered cubic structure
FEB	Focused Electron Beam
FEBID	Focused Electron Beam Induced Deposition
FEG	Field Emission Gun

FIB	Focused Ion Beam
FIBID	Focused Ion Beam Induced Deposition
FM	Ferromagnetic
FWHM	Full Width at Half Maximum
GIS	Gas Injection System
GMR	Giant Magnetoresistance
GPHE	Giant Planar Hall effect
HDD	Hard Disk Drive
IT	Information Technology
LG	Longitudinal geometry
LMIS	Liquid Metal Ion Source
MBE	Molecular Beam Epitaxy
MBJ	Mechanical Break Junctions
MIT	Metal Insulator Transition
MOKE	Magneto Optical Kerr Effect
MR	Magnetoresistance
MIT	Metal Insulator Transition
MOKE	Magneto Optical Kerr Effect
MR	Magnetoresistance
MRAM	Magnetic Random Access Memory
MTJ	Magnetic Tunnel Junction
NP	Nanoparticle
NW	Nanowire
OHE	Ordinary Hall effect
PE	Primary electrons
PECVD	Plasma Enhanced Chemical Vapor Deposition
PG	Perpendicular Geometry
PHE	Planar Hall effect
PLD	Pulsed Laser Deposition
PLR	Precursor Limited Regime
PPMS	Physical Properties Measurement System
RAM	Random Access Memory
RRR	Residual Resistivity Ratio
SCs	Superconductor Materials
SE	Secondary Electrons
SEM	Scanning Electron Microscope
SP	Superparamagnetism
SQUID	Superconducting Quantum Interference Device
SRIM	Stopping & Range of Ions in Matter
STEM	Scanning Transmission Electron Microscopy
STM	Scanning Tunneling Microscopy
STS	Scanning Tunneling Spectroscopy

TAMR	Tunneling Anisotropic Magnetoresistance
TEM	Transmission Electron Microscopy
TG	Transversal Geometry
TMR	Tunneling Magnetoresistance
VRH	Variable Range Hoping
XPS	X-Ray Photoelectron Spectroscopy
XRD	X-Ray Diffraction

Chapter 1
Introduction

This chapter is devoted to explaining the motivation for the study of the nano-structures in this thesis: Fe_3O_4 (half-metal) thin films, metallic (magnetic) atomic-sized constrictions and nanowires based on Pt–C, superconducting W and magnetic Co created by Focused Electron/Ion beams.

The concepts of nanotechnology and nanoelectronics are introduced, setting out the existing limits in current, silicon-based information technology. The break-through represented by the discovery of Giant-Magnetoresistance for information storage in hard-disk-drives is shown as an example of how nanoelectronics can overcome these limits.

A broad description of the nanostructures studied in this thesis and the main objectives of the study are shown. The Dual Beam system used for the nanolithography processes is also detailed.

1.1 Introduction to Nanotechnology

Richard Feynman, Nobel Prize in Physics in 1965, is considered to be the father of Nanotechnology. On December, 29th 1959, he gave a talk at the annual meeting of the American Physical Society, at the California Institute of Technology (Caltech), called "There's Plenty of Room at the Bottom" [1]. In this famous lecture, he opened a new field of science, foreseeing a vast number of applications when manipulating the matter at the atomic and molecular scale.

> I would like to describe a field, in which little has been done, but in which an enormous amount can be done in principle. [...] it would have an enormous number of technical applications. [...] What I want to talk about is the problem of manipulating and controlling things on a small scale. [...] Why cannot we write the entire 24 volumes of the Encyclopedia Britannica on the head of a pin?...

A. Fernandez-Pacheco, *Studies of Nanoconstrictions, Nanowires and Fe₃O₄ Thin Films*, Springer Theses, DOI: 10.1007/978-3-642-15801-8_1,
© Springer-Verlag Berlin Heidelberg 2011

Nanotechnology and nanoscience could be defined as the study and control of matter in structures between 1 and 100 nm (1 nm $= 10^{-9}$ m, about ten times the size of an atom). It is a multidisciplinary field, involving areas such as physics, chemistry, materials science, engineering or medicine. During the following decades, Nanotechnology is expected to have a large influence in our society. Some voices even consider Nanotechnology as a new industrial revolution [2]. Science and technology research in nanotechnology promises breakthroughs in areas such as materials and manufacturing, nanoelectronics, medicine and healthcare, energy, biotechnology, information technology, and national security.

The reduction in size of materials implies that any of their dimensions become comparable to fundamental lengths governing the physics of the system. Therefore, classical laws are no longer suitable for explaining many phenomena as we enter the field of mesoscopic physics, and in some cases the use of quantum physics is necessary.

Materials can be classified taking into account the number of dimensions which are macroscopic, with the rest being nanoscopic: three (bulk material), two (thin films), one (nanowires and nanotubes), zero (nanoparticles and quantum dots). The fabrication and characterization of nanometric structures requires the use of specific sophisticated instruments, with two main strategies for their fabrication. The top-down approximation is based on the use of micro and nanolithography techniques to define nano-objects, starting from bigger structures. The bottom-up approach, however, consists of the fabrication of complex structures by assembling single atoms and molecules into supramolecular entities (see Fig. 1.1). Nowadays, fabrication of integrated circuits is based on the first approach, although applications of this second strategy are expected to reach the market within two or three decades [3].

This thesis is devoted, in a general form, to the creation of micro and nanostructures using top-down methods, for possible applications in nanoelectronics and information storage, with special emphasis on magnetic nanostructures for applications in spintronics.

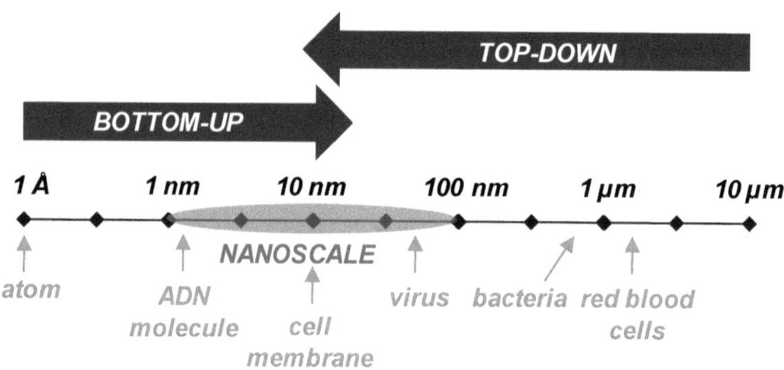

Fig. 1.1 Scheme illustrating the scale-range where nanoscience applies, including the two types of philosophies for fabrication. Some natural entities are included, for didactic comparison

1.2 Introduction to Nanoelectronics

1.2.1 Information Technology: Current Limits

The tremendous impact of information technology (IT) has become possible because of the progressive downscaling of integrated circuits and storage devices. The IT revolution is based on an "exponential" rate of technological progress. The paradigmatic example is computer integration, where Moore's famous (empirical) law [4] predicts that the number of silicon transistors in an integrated circuit will double roughly every 18 months (see Fig. 1.2).

In a computer, the two key aspects to take into account are its ability to process information (logics), and to store/read such information (memory). The transistor is the basic component of the microprocessor (CPU), responsible for performing the logical operations, as well as of the Random-Access-Memories (RAM), the fast and volatile (information is lost when power is off) type of memory in a computer. A CPU chip, composed of a large number of transistors, encloses about 10^9 transistors within an area of 1cm^2 [6]. This high density is obtained by the use of projection optical lithography processes, which are able to overcome the diffraction limit of light [7, 8]. In fact, the transistors are currently fabricated using the so called 45 nm-technology (the name of the technology means that the channel joining the source and drain in transistor has this size. In this case 45 nm). This spectacular downscaling is however, reaching its technological limits [9]. As the limits for the current technology are approaching, the continuation of the IT

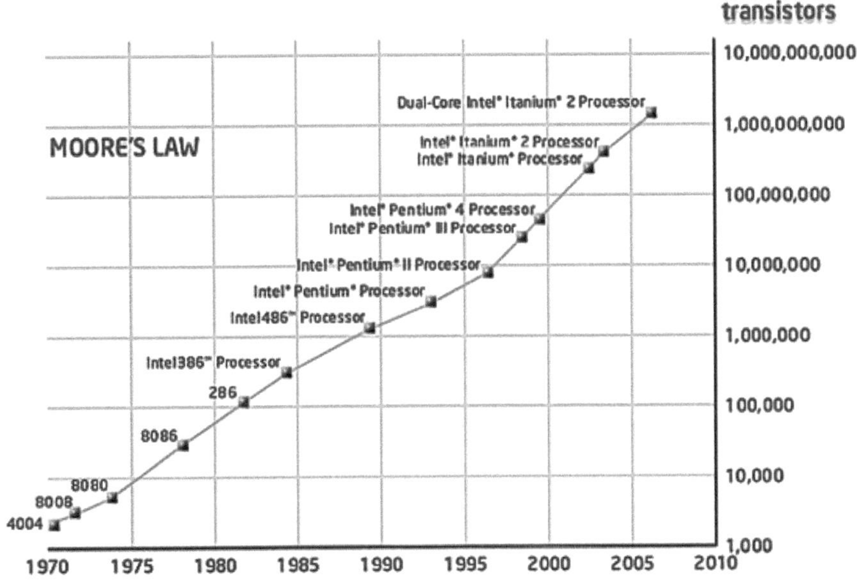

Fig. 1.2 Moore's law. Image taken from Ref. [5]

revolution must be supported by new ideas for information storage or processing, leading to future applications. New technologies should result in greater integration, together with an operation speed of hundreds of GHz, or THz. Three different routes or strategies are today established to improve current electronics [9]:

i. The *More Moore* approach: the same electronic components as currently used are decreased in size. This is also based on designing more efficient electronic components and better system architectures.
ii. The *More than Moore* approach: using the new functionalities appearing as a consequence of the gradual integration of CMOS technology.
iii. The *Beyond Moore* approach: based on the design of new system architectures, whose building blocks may be atomic and molecular devices. Besides, boolean logic is substituted by novel algorithms, possibly based on quantum computation.

Moore's law is an example of a general trend for IT areas. Thus, internet traffic doubles every 6 months, wireless capacity doubles every 9 months, optical capacity doubles every 12 months, magnetic information storage doubles every 15 months, etc. [10].

1.2.2 GMR Heads: Impact on the Information Storage

To illustrate how nanoelectronics can overcome the limits of current technology, we will show in this introduction one of the first examples of fundamental physical concepts in the nanoscale representing a technological breakthrough, and being incorporated in the mass market production with widespread applications: the discovery of the Giant Magnetoresistance (GMR) effect, implying the implementation of new read heads for magnetic hard-disk drives (HDDs).

As an initial brief description, HDDs are non-volatile memory devices, storing binary information using magnetic materials. They consist of three components (see Fig. 1.3). The first is the *storage medium*, which is the disk in which the data is actually stored in the form of small magnetic areas. The second is the *write heads*, consisting of a wire coil wound around a magnetic material, which generates a magnetic field (by electromagnetic induction) when a current flows through the coil. The magnetic field writes the data by magnetizing the small data bits in the medium. Finally, the *read heads* sense the recorded magnetized areas. We will go into detail about this third component only, in order to illustrate the impact of nanotechnology (for further information about the other two see, for instance, Refs. [11, 12]).

In Fig. 1.4, the evolution of areal density of magnetic HDDs during the recent decades is shown. The introduction in 1997 by IBM of the Giant Magnetoresistance (GMR) sensor to replace the previously used Anisotropic-Magnetoresistance (AMR) sensor meant that the growth rate for storage immediately increased up to 100% per year. By providing a sensitive and scalable read technique, the HDD

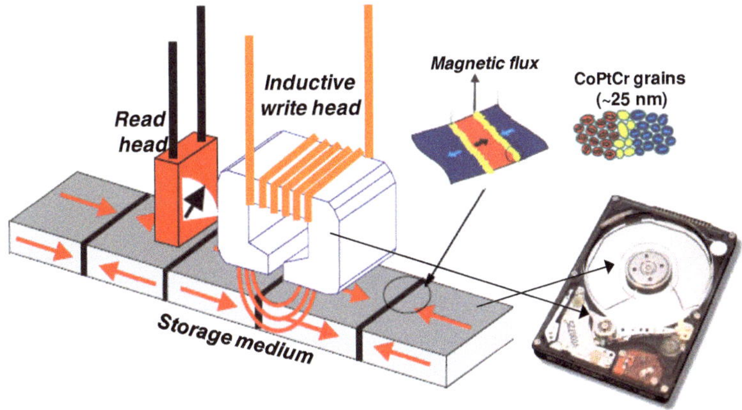

Fig. 1.3 Magnetic hard disk drive for information storage. Bits are stored in nanometric CoPtCr grains, where "1" and "0" are defined by the direction of the magnetization in grains. The information is written by inductive heads, which create a magnetic field. Read heads were initially inductive, and currently based on magnetoresistance effects. The bottom-right image shows a real HDD. Images adapted from Ref. [13]

areal recording density increased by three orders of magnitude (from ~ 0.1 to ~ 100 Gbit/in.^{2}) between 1991 and 2003[9].

We will briefly discuss the physical origin of GMR. Its discovery should be set in the context of the time, the end of the 1980s, when an excellent control on the growth of thin films was acquired, permitting the fabrication of metallic (magnetic) superlattices and heterostructures. A fascinating outcome of this research was the observation of interlayer coupling between ferromagnetic (FM) layers separated by a non-magnetic metallic (M) spacer [14]. This coupling oscillates from FM to antiferromagnetic (AF) with an increase in the spacer layer

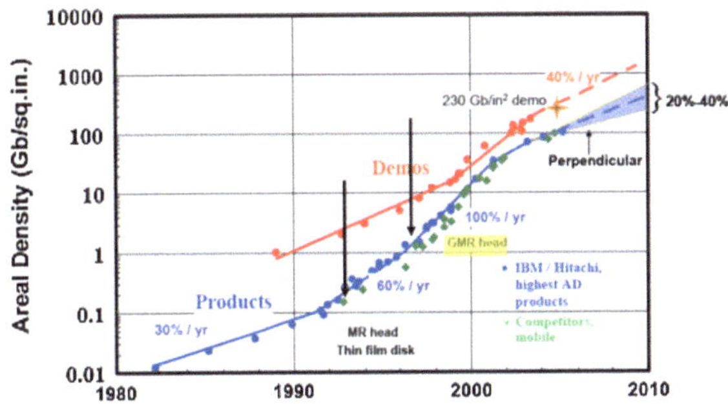

Fig. 1.4 Evolution of HDD areal density in recent decades. In 1991, the introduction of AMR-based reads produced a growth in the areal density up to 60% per year. In 1997, the substitution of AMR-based read heads by GMR heads implied an immediate growth rate increase, from 60% up to 100% per year. Further strategies are perpendicular storage and tunneling magnetoresistance read heads. Image taken from Ref. [13]

thickness, described by the RKKY interaction [15]. When the FM layer thickness becomes comparable to the mean free path of electrons, spin-dependent transport effects appear. In Fig. 1.5a, the simplest model of the GMR effect is schematically shown, where electrons are conducted through alternating FM and M materials.

The two-current model proposed by Mott [16] and applied in FM metals by Fert and Campbell [17] establishes that at low temperatures spin-up and spin-down electrons are conducted in two separated channels, as a consequence of the different spin-dependent scattering probabilities. Electrons passing though the first FM become spin polarized. If the next M layer is thin enough, this polarization is maintained. When these electrons flow to the second FM, two possible resistances can result, depending on the relative orientation of the FM layers (see Fig. 1.5a). For applications, devices called spin-valves are constructed (see Fig. 1.5b), where these effects are up to 20% at 10–20 Oe [18]. The simple concept behind spin valves consists of fixing one FM layer along one orientation (normally by the AF interaction with adjacent layers), and leaving the other one free. This free layer "opens" or "closes" the flow of electrons. Two types of geometries are possible, the current-in-plane geometry (CIP), used today in sensors, and the

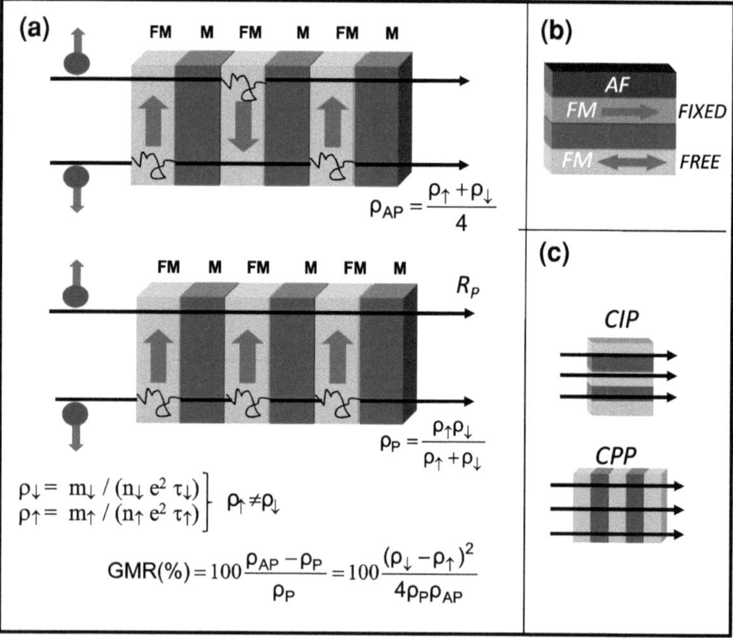

Fig. 1.5 a Scheme of the GMR effect. Spin-up and down electrons are conducted in two separated channels. As a consequence, the parallel configuration is less resistive than the antiparallel one (see text for details). **b** Scheme of the simplest spin-valve device, formed by a fixed-FM (*top*), an M, and a free-FM (*bottom*). The magnetization direction of the top layer is pinned by the exchange-bias coupling to the adjacent AF layer. **c** Two geometries for the GMR. The CPP gives higher MR effects than CIP

current-out-of-plane (CPP) (see Fig. 1.5c). The underlying GMR physical effects are more complex than this simple explanation suggests. The introduction of intrinsic potentials, models in the ballistic regime of conduction, or the introduction of the spin-accumulation concept are sometimes used. For details see, for instance, Refs. [18–23].

The GMR was discovered almost simultaneously by Fert's [24] and Grünberg's [25] groups and, due to its huge influence in IT with applications in magnetic storage, sensors and electronics [21], it resulted in the award of the 2007 Nobel Prize for Physics for both scientists. GMR is the founding effect for spintronics. In conventional electronic devices, the charge is the property of the electron which is exploited. In magnetic recording, the spin (through its macroscopic manifestation, magnetization) is used. In spintronics, effects where both charge and spin and their interplay are studied, and are manifested in nanoscale magnetic materials. Spintronics is one of today's main research lines in nanoelectronics.

1.3 Comparison of the Nanostructures Studied in the Thesis from a Dimensional Point of View

We have basically studied three types of nanostructures in this thesis, taking geometrical considerations into account (Fig. 1.6):

(1) 2D-thin films (of Fe_3O_4: Chap. 3): nanometric in one dimension, macroscopic in the other two.
(2) 1D-nanowires (of Pt–C: Chap. 5, of W–C: Chap. 6, of Co: Chap. 7): nanometric in two dimensions (thickness and width), macroscopic in one (length).
(3) 0D-atomic nanoconstrictions (of Cr and Fe: Chap. 4): macroscopic electrodes connected through a few atoms.

The fabrication of thin films makes necessary the use of growth techniques which have a good control of the deposition processes, such as Molecular Beam Epitaxy (MBE), Pulsed Laser Deposition (PLD), Chemical Vapor Deposition (CVD), etc. To create nanowires or nanoconstrictions, nanolithography tools are also required to define the lateral nanometric sizes. In our case we have used a Dual Beam system (see Sect. 1.5).

We can make a rough comparison of how many atoms are contained in a $10\ \mu m \times 10\ \mu m \times 10\ \mu m$ cube for all the geometries considered, by assuming a typical size of 0.1 nm for an atom, and using common sizes of the materials studied. Thus, whereas a bulk material would contain 10^{15} atoms, a thin film (thickness $=$ 20 nm) would have 2×10^{12} atoms. A nanowire (thickness $=$ 50 nm, width $=$ 100 nm) would contain 5×10^{10} atoms. For nanoconstrictions, although we would calculate a value of the order of 10^{13} atoms (and their response would be measured by a macroscopic probe), the electrical conduction is dominated by a few tens of atoms, those forming the constriction and a few neighbors in the electrodes. Furthermore, in this case, superficial atoms are approximately 100%, whereas they are of the order of

Fig. 1.6 Scheme of the different nanostructures studied in the thesis, for comparison of their dimensions. Thin film is thickness-nanometric, nanowires are thickness-and-width-nanometric, whereas an atomic constriction is formed by bulk electrodes joined by a few atoms

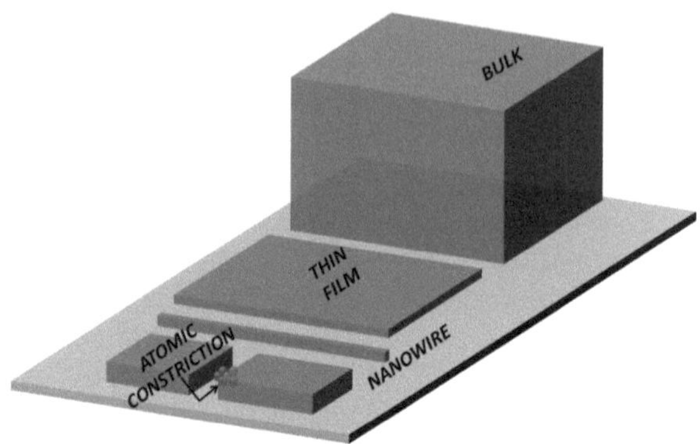

$\sim 10^{-3}\%$ in the other geometries. The scaling shown when reducing dimensions is still more spectacular if nanometric dimensions approach the nanometer. Surface effects, defects, confinement effects, etc. gain importance versus intrinsic properties as a smaller amount of atoms plays a role in the phenomenon studied. Some of the effects studied in this thesis are, in some way, manifestations of this.

In the next sections we describe in more detail the motivation for the study of these nanostructures, as well as the capabilities of the tool used (a Dual Beam system) for the nanolithography processes.

1.4 Epitaxial Fe$_3$O$_4$ Thin Films

In Sect. 3.1 a detailed description of the structural, magnetic and conduction properties of Fe$_3$O$_4$ is given. Here we only focus on the interest of this material for spintronics.

In a solid, the spin polarization P can be defined as.

$$P = \frac{N_\uparrow(E_F) - N_\downarrow(E_F)}{N_\uparrow(E_F) + N_\downarrow(E_F)} \tag{1.1}$$

where $N_\uparrow$ and $N_\downarrow$ are the density of states (DOS) with up and down spins, respectively. Spin polarization is, then, a measurement of the net spin of the current flowing through a material. If only one type of electron is responsible for conduction, $|P| = 1$, the material is referred to as "half-metal". In Fig. 1.7 a scheme of bands in a non-magnetic, a normal ferromagnetic (such as iron, for example) and a half-metal material is shown.

In the 1980s band calculations [26, 27] predicted that the conduction in magnetite above T_v ($T_v = 120$ K: Verwey temperature, see Sect. 3.1 for details) is fully polarized, with $P = -1$. In practice, photoemission [28, 29], tunneling magnetoresistance (MR) [30] or STM [31] measurements have evidenced a high negative spin polarization. Apart from magnetite, other compounds have been

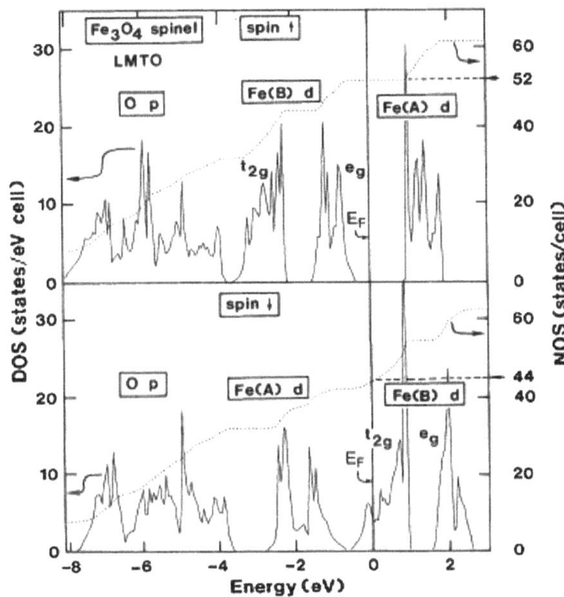

Fig. 1.7 Bands scheme of a non-magnetic, a ferromagnetic and a half-metal ferromagnetic material. In the half-metal, all the electrons responsible for conduction (those at the Fermi level) have only one spin direction

Fig. 1.8 Theoretical calculation of the DOS in Fe₃O₄. Taken from Ref. [27]

predicted as half-metals [32, 33]: CrO_2, manganites, double-perovskites and heusler alloys (Fig. 1.8).

An enormous technological interest exists in studying half metal materials, as magnetic electrodes for magnetic tunnel junctions (MTJs). MTJs are devices similar to those exhibiting GMR (CPP configuration, see Sect. 1.2.2), but instead of metallic spacers separating FM layers, a very thin non-magnetic insulator, I (typically 1–3 nm) is used as a barrier (see Fig. 1.9a). Electrons pass from one FM to the other by the tunnel effect, which conserves the spin. A high or low resistance state defines a magnetoresistance (tunneling magnetoresistance, TMR) as we previously explained in the case of GMR. Considering the model developed by Jullière [34] for the TMR:

$$\text{TMR } (\%) = 100\frac{2P_1P_2}{1 - P_1P_2} \qquad (1.2)$$

The TMR will diverge if both FM electrodes are half metallic ($P_1 = P_2 = 1$). We should, however, remark that in this simple model the DOS of the electrodes alone defines the tunneling current, but in reality the insulator barrier is also fundamental for the TMR signal [36] since it can act as a symmetry filter for the wave functions of tunneling electrons [37, 38].

Since the observation of a very high TMR at room temperature by Moodera et al. [39] in 1995, a great deal of research has been devoted to this new effect. TMR values much higher than GMR can be obtained, with record values at low temperature of a few hundreds percent, using MgO barriers and half-metal electrodes [40].

From the point of view of applications, MTJs with TMR $\sim 50\%$ are used. The higher sensitivity of TMR versus GMR makes it very appealing for applications. TMR read heads have indeed been commercialised since 2005 [41], substituting the spin valve heads in HDDs. These high values have resulted in a use for MTJs, not only for sensing, but also as another type of fast-access non-volatile memory the Magnetic-Random-Access-Memory (MRAM), see Fig. 1.9c. However, the future of MTJ-based technology is now uncertain, since MTJs have intrinsic high resistances, making it impossible to maintain current high signal-to-noise ratios

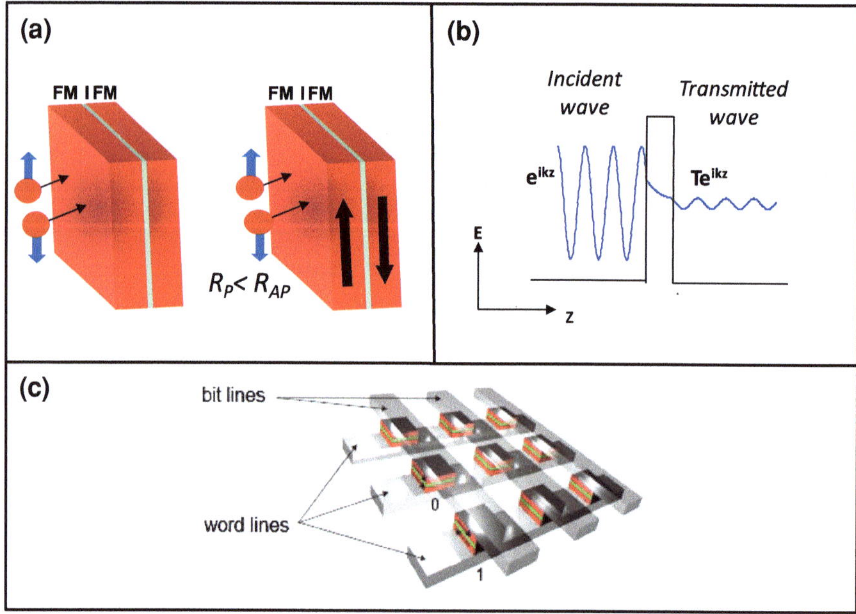

Fig. 1.9 Magnetic tunnel junction. **a** Scheme for the low (parallel electrodes) and high (antiparallel electrodes) resistance states. **b** Tunnel effect representation, where the evanescent wave of an electron tunnels through a square barrier. **c** MRAM architecture (image taken from Ref. [35]). Bits are stored in the MTJs (0: P; 1: AP). Information is read by measuring the TMR value, whereas it is written, either by the magnetic field generated by the flow of current though the two read lines, or by a spin-transfer effect

with further downscaling. CPP spin valve-memories are nowadays proposed as an alternative [18].

In summary, Fe$_3$O$_4$ is a very promising material to form MTJ electrodes: it has a high ferromagnetic Curie temperature ($T_c = 860$ K), a relatively high saturation magnetization ($M_s = 480$ emu/cm^3) and a high spin polarization at room temperature. Fe$_3$O$_4$ films can be grown epitaxially on MgO (001) substrates, due to the matching lattice parameters ($a_{Fe_3O_4} = 8.397$ Å $\approx 2 \times a_{MgO}$). Before its implementation in real devices, a good control of the growth, magnetic and transport properties in the form of highly crystalline thin films should be achieved. Epitaxial Fe$_3$O$_4$ films have structural defects not present in the bulk material, called anti-phase boundaries, which have a relevant influence on the properties of the films. A systematic study of the magnetotrasport properties of Fe$_3$O$_4$ thin films has been one main objective of this thesis (Chap. 3).

1.5 Dual Beam System for the Fabrication of Nanostructures

To define structures with nanometric lateral sizes, we have used a "dual beam" system. A dual beam incorporates a focused ion beam (FIB) column and a scanning electron microscope (SEM) column in a single system. This combination makes this equipment a very powerful nanolithography tool, with important advantages over a single-beam SEM or FIB system. The typical dual-beam column configuration is a vertical electron column with a tilted ion column. Figure 1.10 shows such a configuration with the ion beam at 52° tilt to the vertical direction. The sample will be tilted to 0° and 52° in such a way that the electron or ion beam is, respectively, perpendicular to the substrate plane. To enable ion milling and electron imaging of the same region, dual beams typically have a coincident point where the two beams intersect with the sample. This is the normal operating position for the system.

We will now give an overview of some key aspects of the system and the nanolithography processes performed.

Fig. 1.10 Scheme of the Dual Beam configuration. SEM and FIB column form an angle of 52°. Both beams are focused on a coincidence point. The stage rotates depending on the process

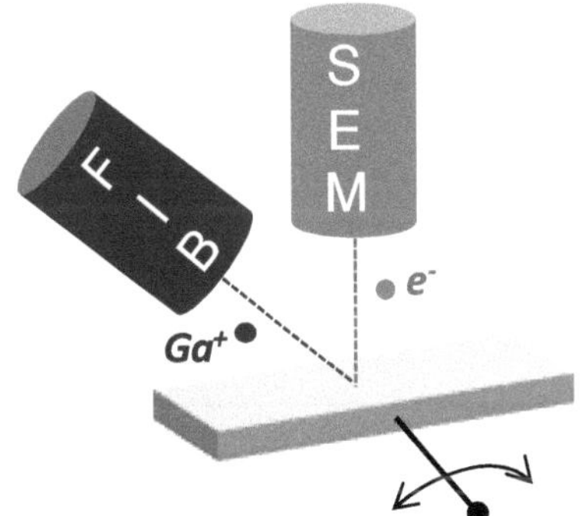

1.5.1 Focused Electron Beam (or SEM)

The FEB column used is a Schottky Field-Emission-Gun source (FEG), where electrons are emitted from a sharp-pointed field emitter, at several kilovolts negative potential relative to a nearby electrode, causing field electron emission. Schottky emitters are made by coating a tungsten tip with a layer of zirconium oxide, which increases its electrical conductivity at high temperature. The FEG produces an electron beam smaller in diameter, more coherent and with up to three orders of magnitude greater brightness than conventional thermionic emitters such as tungsten filaments. The result is a significantly improved signal-to-noise ratio and spatial resolution.

Electron optics share similar characteristics to those of light optics. The FEB column typically has two lenses. The condenser lens is the probe forming lens, and the objective lens is used to focus the beam of electrons at the sample surface. A set of apertures of various diameters define the electron current on the sample, and determine the probe size. Big apertures give high currents and low spatial resolution. For SEMs the acceleration beam energy is usually in the 1–30 keV range and the minimum beam diameter is near 1 nm.

The beam is scanned along the sample, using electro-magnetic lenses. As the beam is rastered, particles are emitted from the surface, reaching the detectors. The convolution of that signal forms an image.

As a consequence of the interaction of primary electrons (PE) with matter, both electrons and photons are emitted by the specimen surface. In Fig. 1.11a we show those used in our equipment to obtain information. In this section we will only focus on two main types of electrons generated (Auger electrons, X-rays and cathodoluminescence will not be commented on here):

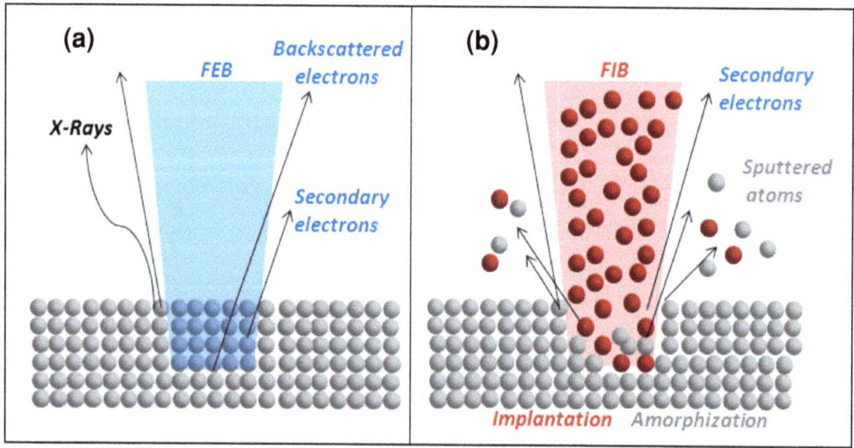

Fig. 1.11 Interactions between a focused electron (**a**) or ion (**b**) beam and a substrate. The main effects in both cases are schematically shown

- *Secondary electrons (SE):* These electrons are produced due to inelastic collisions with weakly bound conduction band electrons of metals. Although they are emitted along the whole trajectory of the PE, only those close to the surface are collected, since their energy is only of a few eV.
- *Backscattered electrons (BSE)*: These are elastically reflected primary electrons, with energies close to the PE, being therefore emitted from a large volume.

1.5.2 Focused Ion Beam

The FIB apparatus is very similar to a SEM. It is composed of three main parts: the source, the ion optics column, and the stage and beam control. The FIB optics and stage and beam control are very similar to those of the FEB, explained in the previous section. The ion source is a liquid metal source (LMIS), consisting of a reservoir of liquid gallium, which feeds the liquid metal to a sharpened needle, made of tungsten. The LMIS has the ability to provide a source of ions of ~ 5 nm in diameter [42], with a high degree of brightness, typically $\sim 10^6$ A/cm^2 sr. The heated Ga flows and wets the W needle. An electric field applied to the end of the wetted tip causes the liquid Ga to form a point source in the order of 2–5 nm in diameter at the sample surface. The extraction voltage can pull Ga from the W tip and efficiently ionize it. Once the ions are extracted from the LMIS, they are accelerated through a potential (5–30 kV) down the ion column.

Figure 1.11b shows a schematic diagram illustrating some of the possible ion beam/material interactions that can result from ion bombardment of a solid. When the ions collide elastically with the target, atoms are sputtered if the energy supplied is higher than the binding energy of the material. Inelastic interactions also occur, producing, among others, the emission of SE. Detection of SE is the standard imaging method in the FIB. Others effects such as amorphization or redeposition are also inherent to a FIB process. For details see Refs. [43, 44].

1.5.3 Focused Electron/Ion Beam Induced Deposition

Anyone using an SEM will probably have observed that the surface becomes contaminated when the beam exposes the same place for some time. This film is formed from remains of hydrocarbons present in the vacuum system. This undesired effect can, however, be used to deposit a material locally, if the appropriate gas is adhered to the surface. This technique is a major application of the Dual Beam system, called Focused-Electron-Beam-Induced-Deposition (FEBID) and Focused-Ion-Beam-Induced-Deposition (FIBID), depending on the beam used (other names are used in the literature for these techniques: EBID/IBID, EBICVD/IBICVD).

FEBID and FIBID involve a chemical vapor deposition process that is assisted by electron and ion beams respectively (for recent review articles, see Refs. [44–46]).

The basic principle is simple: a Gas-Injection-System needle (GIS) is inserted near the substrate. Gas is injected into the chamber, becoming adsorbed on the surface. The beam electrons/ions interact with these gas molecules, decomposing them. As a consequence, the volatile fragments are evacuated from the vacuum chamber, the rest being deposited. Thus, the creation of patterns with nanometer resolution is performed in a direct way, just as "a pencil writing on a paper". This simplicity is the main advantage with respect to other more well-established lithographic techniques, where several steps, usually involving resists, are required.

Going into more detail, three main physical–chemical processes should be considered [46]:

– The first is the substrate–precursor molecule interactions: mechanisms such as diffusion, adsorption and desorption have to be taken into account.
– The second is the electron/ion-substrate interaction: a primary beam of electrons/ions is focused onto the substrate. Some of these electrons/ions collide and are deflected from their initial trajectory whereas others undergo inelastic collisions transferring the energy to other electrons/ions.
– The third is the electron/ion-precursor molecule interaction: the scission of a bond in a precursor molecule by an electron or ion occurs when the generated electrons have energies of a few eV. (Fig. 1.12).

Three deposition regimes have been identified [44–46], which depend on all the factors described above:

– *Electron/ion-limited regime*: The growth is limited by the current density and is independent of the gas flux.
– *Precursor-limited regime*: The growth is limited by the number of molecules, coming from the GIS, and arriving at the irradiated area.
– *Diffusion-enhanced regime*: The supply of gas molecules for deposition is dominated by the surface diffusion.

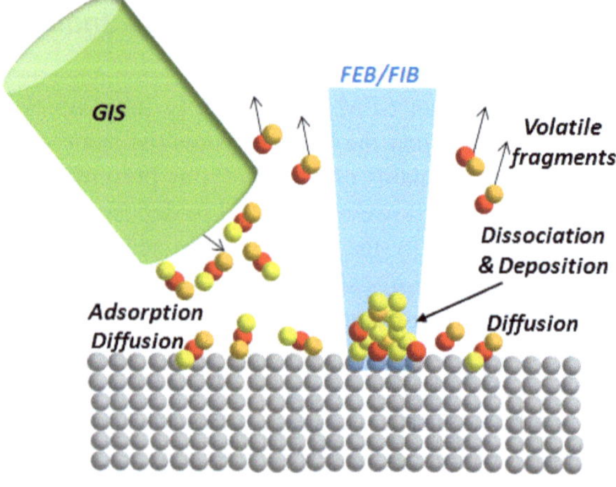

Fig. 1.12 Scheme for FEBID and FIBID. Molecules are injected by the GIS, become adsorbed and then diffused at the surface. Some are decomposed. The non-volatile fragments are deposited, whereas the volatile fragments are pumped away

The physical phenomena involved in the deposition are complex and models trying to explain experimental results take into account factors such as the electron flux as well as their spatial and energy distribution, the cross section of the precursor as a function of electron energy, the precursor residence time, the electron-beam induced heating, etc. An important piece of information has been gained through Monte Carlo simulations [44–46]. These simulations have highlighted the important role played by the generated secondary electrons in the deposition rate and also in the spatial distribution and resolution of the nanodeposits.

A typical situation for these techniques is to use an organo-metallic gas precursor for local deposition. In general, these gases are not fully decomposed, and a carbonaceous matrix is also present in the deposits. The typical metal content for deposits by FIBID is around 30–40% atomic, with the rest being mainly carbon, whereas for FEBID the metal content is usually only around 15–20%. Thus, the main drawback when creating metallic nanostructures with these techniques is the high carbon percentage present in the deposits, which can influence dramatically the properties of these nanostructures. Post-growth purification processes are sometimes used to increase the metallic content [47, 48]. In the case of FIBID, the damage induced by ions is also an important issue to take into account. These aspects, together with a higher deposition yield for FIBID with respect to FEBID, are the most relevant differences between depositing with one or other column.

FEBID and FIBID allow local deposition in the targeted place with controllable lateral size, determined by the volume interaction of the beam with the substrate (minimum size from 15 nm to 10 µm) and thickness (typically ranging from 10 nm to 10 µm). Amongst the applications of FEBID and FIBID deposits, one can cite the reparation of optical masks and integrated circuits [49], fabrication of three-dimensional nanostructures [50], deposition of protection layers for lamella preparation [42], creation of electronic nanodevices [51], fabrication of nano-electrodes and nanocontacts [52], transport studies of nanowires [53], deposition of magnetic [54] or superconductor materials [55], etc. The technique possesses unique advantages for fabricating nanostructures compared with more conventional techniques (such as Electron Beam Lithography, EBL, for instance). It is a one step mask-less process, with nanometric resolution. Sophisticated structures can be routinely created, in principle, on any surface.

In order to illustrate the advantages over other nanolithography techniques better established in the scientific community, we show in Fig. 1.13 a comparison between EBL, FIB and FEBID/FIBID. The simplicity and flexibility of using a Dual Beam for nanolithography is evidenced, since typical methods used nowadays such as EBL consist of several critical steps, using resists for nanopatterning.

In this thesis we have used three precursors for the growth of nanowires, which are shown in Fig. 1.14: methylcyclopentadienyl-trimethyl platinum [$(CH_3)_3Pt(CpCH_3)$], tungsten hexacarbonyl [$W(CO)_6$] and dicobalt octacarbonyl [$Co_2(CO)_8$]. The decomposition of these precursors is very different, resulting in completely different physical properties.

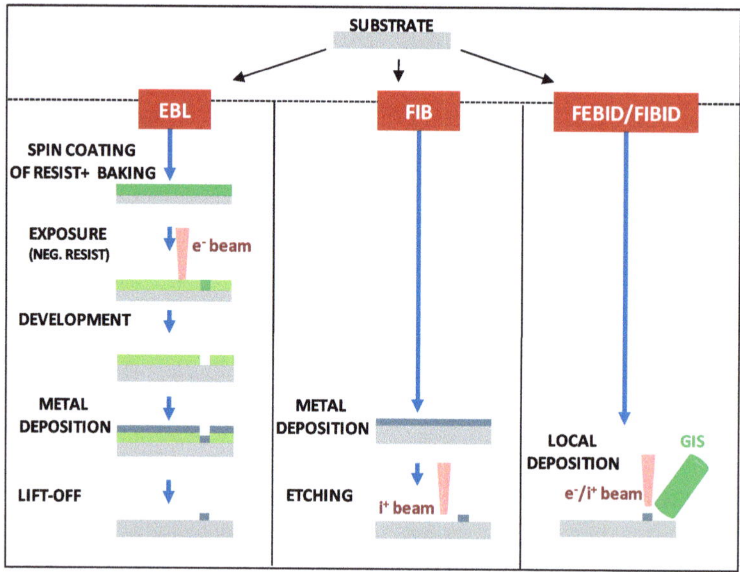

Fig. 1.13 Scheme comparing an EBL (lift-off) a FIB and a FEBID/FIBID process

Fig. 1.14 Chemical structure of the three metal–organic gases used in this thesis. **a** $(CH_3)_3Pt(CpCH_3)$. **b** $W(CO)_6$. **c** $Co_2(CO)_8$. All of them are stored in solid form, being heated up to the sublimation point for their use

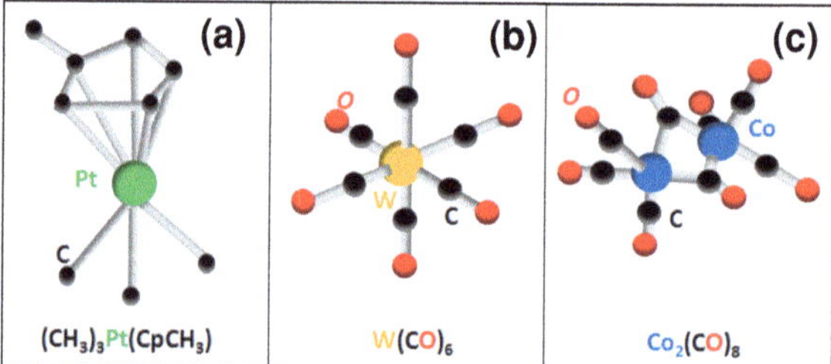

1.6 Atomic-Sized Nanoconstrictions

1.6.1 Theoretical Background for Atomic-Sized Constrictions

1.6.1.1 Introduction

In the macroscopic world, the nature of electrical conduction in metals does not depend on their size. However, when this size approaches the atomic regime, electrons can be conducted without being scattered, and Ohm's Law is no longer fulfilled. In this regime, known as ballistic, the normal concept of conduction is changed, it being necessary to invoke the wave nature of the electrons in the conductor to describe it. Quantum effects become visible at room temperature due to the large energy scales present [56].

Fig. 1.15 Schematic illustration of a diffusive (*left*) and ballistic (*right*) conductor. Adapted from Ref. [56]

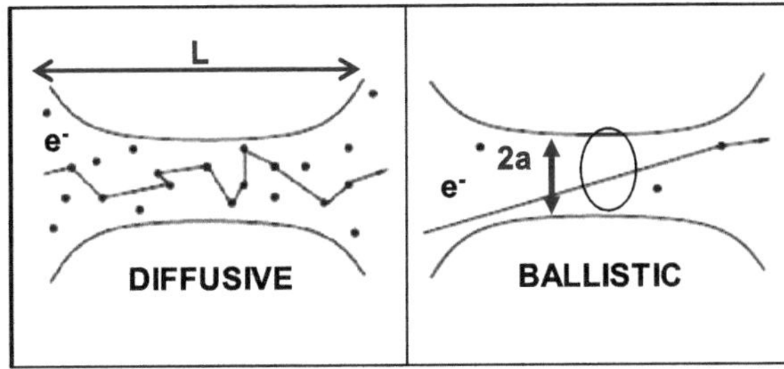

The creation of constrictions with these dimensions is a hard task, and laboratory techniques which operate at the molecular and atomic scale are necessary, normally under very strict environmental conditions.

1.6.1.2 Conduction Regimes for Metals

Let us imagine two big metallic electrodes, connected by a constriction (also denoted as contact) of length L, through which electrons flow.

If this contact has macroscopic dimensions, the electrons are scattered when passing (Fig. 1.15, left), since $L > l$ (where l is the elastic mean free path, typically a few tens of nanometers in metals at low temperatures). In this case, Ohm's law is fulfilled, and the conductance of this constriction is given by the classical expression:

$$G = \sigma S/L \tag{1.3}$$

where σ is the conductivity of the material, and S is the section of the contact. We call this regime of conduction *diffusive*.

When the dimensions of a contact are much smaller than l the electrons will pass through it without being scattered (Fig. 1.15, right). This regime is called *ballistic*. In such contacts there will be a large potential gradient near the contact, causing the electrons to accelerate within a short distance. This problem was first treated semiclassically by Sharvin [57], obtaining for the conductance:

$$G = \frac{2e^2}{h}\left(\frac{\pi a}{\lambda_F}\right)^2 \tag{1.4}$$

where a is the radius of the contact. In this case, the resistance is not proportional, as was the case in the diffusive regime, to the number of scattering events. The resistance of the ballistic channel tends to zero, and it is the boundaries between the leads and channel which are responsible for the resistance of the system. In Sharvin's formula, G depends on the electronic density through the Fermi wavelength λ_F, and it is independent of σ and l. Quantum mechanics are only introduced in this treatment by Fermi Statistics [56].

This semiclassical approach is not, however, valid when the contacts are atomic-sized, with a few atoms contributing to the conduction of electrons. In this case L $\sim \lambda_F$ (from a few to fractions of Å) and we enter into the *full quantum limit*. This problem was treated by Landauer [58], reaching a value for the conductance of:

$$G = \frac{2e^2}{h} \sum_{i=1}^{N} \tau_i \qquad (1.5)$$

where τ_i is the transmission coefficient of the channel i. The conductance is proportional to the value.

$$G_0 = \frac{2e^2}{h} \qquad (1.6)$$

which is called the "quantum of conductance", and whose corresponding resistance value, the quantum of resistance, is $R_0 = 12.9$ kΩ. Thus, the wave nature of electrons is manifested by the quantization of the conductance. This result is analogous to the propagation of waves through wave guides, when the channel width is of the order of the incident wavelength.

The presence of electron wavefunctions at the Fermi level determines the number of conduction channels. In a metallic constriction, the conduction is determined on both the valence electrons and transmission factors, which depend on the contact geometry. The discrete jumps observed when approaching the ballistic regime are therefore not associated to the quantization of the conductance, but on the transition between stable configurations as the wire is narrowed [56, 59]

1.6.1.3 Typical Methods for the Fabrication of Atomic Contacts

As was previously commented, the methods for creating atomic contacts are mainly laboratory techniques, normally working at Helium liquid temperature, and in ultra-high vacuum conditions. These experiments have led to huge advances in the understanding of electron conduction in atomic-sized metal structures. Among the most commonly used we should cite:

  (i) Mechanical break junctions (MBJ) [60]: a metallic film is broken in a controllable way, by bending a flexible substrate on which the material has previously been deposited.
 (ii) Use of STM [61]: the tip is contacted to a surface, and while the current is monitored, the distance between the tip and the sample is varied at a constant low bias voltage (~ 10 mV).
(iii) Electrical break junctions (EBJ) [62]: atoms are electromigrated by the application of large current densities on a previously (usually by EBL) patterned thin film.
(iv) Electrochemical junctions (ECJ) [63]: in an electrochemical bath, a previously patterned (EBL) gap is filled by electrodeposition.

1.6.2 Atomic Constrictions in Magnetic Materials

1.6.2.1 Introduction

Due to the spin degeneracy of the conduction electrons in magnetic materials, the quantum of conductance includes a factor of two. The strong exchange energy in magnetic materials produces the splitting of the conduction bands, making the transport spin-dependent. Therefore, the conductance quantum in magnetic materials is thought to be:

$$G_0^M = \frac{e^2}{h} \tag{1.7}$$

1.6.2.2 Ballistic Magnetoresistance

Since the conduction in metals in the ballistic regime is not affected by spin-dependent diffusion, the MR properties of magnetic materials should therefore be governed by the band structure. In the case of GMR, this fact was studied theoretically by Schep et al. [64], although the experimental results in this case were unsuccessful [65].

The topic was taken up again with spectacular results, when MR values of a few hundred percent were measured in Ni contacts formed by approaching wires mechanically [66]. These results were also obtained in nanocontacts of other FM materials by using different techniques, by several groups [67–70]. The effect proposed, coined as Ballistic Magnetoresistance (BMR), was associated to the presence or non-presence of a domain wall in the contact, blocking or letting the electrons pass through the contact (see Fig. 1.16a). When the magnetizations in the electrodes are parallel, the conductance is of the order of the conductance quantum, but in the antiparallel case, electrons are partially reflected because the majority spin channel changes its spin direction in the contact. Hence, MR would be expected to reach extremely large values when the sample resistance is of the order of $R_0 = 12.9$ kΩ.

These results had a large impact on the spintronics community, since new devices based on this effect, higher than GMR, could be performed. However, subsequent experiments [71–73] trying to reproduce previous results revealed that actually mechanical artifacts, such as atomic reconfigurations or magnetostriction

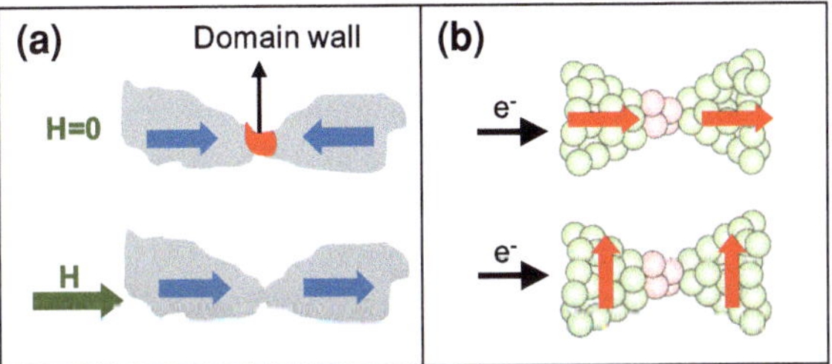

Fig. 1.16 Schematic illustration for the MR effects in atomic-sized magnetic contacts: **a** BMR and **b** BAMR

(change of volume in a magnetic material when its magnetization changes), were responsible for this high effect, and in fact BMR would be only of the order of a few tens percent [74]. Currently, the scientific community has basically concluded that the impressive BMR is a consequence of spurious effects [74], although it should also be remarked that it is still not clear if magnetic contacts with high transmission factors can be fabricated [56].

1.6.2.3 Ballistic Anisotropic Magnetoresistance

Another different finding in atomic-sized contacts was a large anisotropy in the MR of the contacts, whose magnitude and angular dependence were found to be very distinct from bulk materials, the so called ballistic anisotropic magnetoresistance (BAMR) (see Fig. 1.16b).

Bulk FM materials (in the diffusive regime) have a different MR value depending on the relative direction of the current with respect to the magnetization, the so called Anisotropic Magnetoresistance (AMR). Briefly, in a FM it is possible to describe the conductance in two separated channels, for spin up and down directions [16, 17]. When adding to the energy of the system the spin–orbit coupling (L–S), the wavefunctions of electrons have to be recalculated, no longer being eigenstates of S_z, since the L–S mixes spin states. When a magnetic field is applied, the spin magnetization direction rotates with it, the 3d electron cloud deforms, due to the L–S, and changes the amount of scattering of the conduction electrons. This results on an anisotropy of the resistance [75, 76]:

$$\text{AMR} \ (\%) = 100\frac{\rho_{//} - \rho_{\perp}}{\rho_0} \tag{1.8}$$

The sign of the AMR depends on the band structure of the FM, normally being positive in metals ($\rho_{//} > \rho_{\perp}$).

The dependence of the AMR with the angle θ formed by the current and the magnetization in an isotropic material (for example a polycrystal) is:

$$\rho(\theta) - \rho_{\perp} = \left(\rho_{//} - \rho_{\perp}\right)\cos^2\theta \tag{1.9}$$

In the ballistic transport, the AMR has to be recalculated, since the structural, electronic and magnetic properties of atoms change. For instance, the orbital moment will not be as quenched as in bulk [77], and the spin–orbit coupling will increase substantially [74]. Theoretical studies of BAMR in chains of Ni and Co [78, 79] indicate that the effect would be much higher than in bulk, of the order of G_0, and with a much more abrupt angle dependence.

Some experimental measurements support these theories. Specifically, we can cite the work by Sokolov et al. [80], where in Co contacts made by electrode-position, changes in the AMR of the order of the quantum of conductance were found, and the work by Viret et al. [81], where iron contacts fabricated by mechanical break junction were systematically studied, finding even higher effects for samples in the tunneling regime than in the metallic (ballistic) regime.

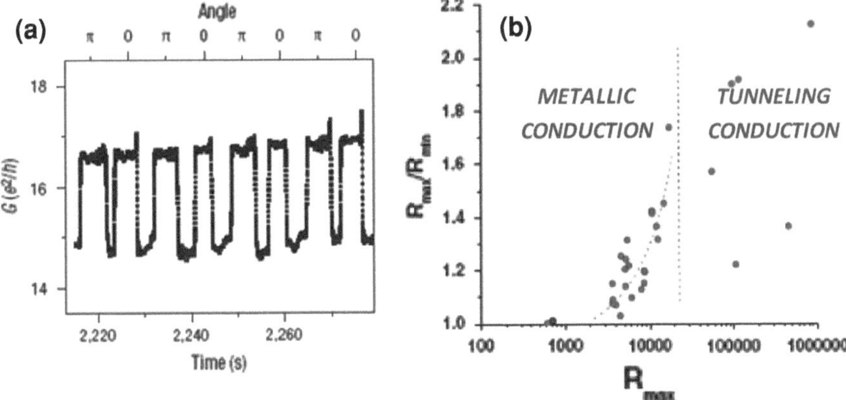

Fig. 1.17 **a** Conductance of a Co ECJ-nanocontact as a function of the out-of-plane magnetic field (1 T), measured at 300 K. Jumps of $\sim 2G_0$ are observed [80]. **b** BAMR magnitude of ECJ-Fe contacts, both in the ballistic and the tunneling regime [81]. Images taken from references cited

However, critical voices have claimed that atomic reconfigurations in the contact, rather than an intrinsic electronic effect, are the most feasible reason for this behavior [82]. Moreover, recent calculations in Ni nanocontacts question the existence of large BAMR [83] (Fig. 1.17).

The objective of this part of the thesis has been the development of a method for the fabrication of atomic-sized constrictions in metal using the FIB. The good adherence of the atoms to the substrate should minimize artifacts, making possible the study of the MR in the ballistic regime of conduction. The implementation of devices based on this effect would be eventually possible, since FIB processes can be introduced into electronics manufacturing protocols.

1.7 Functional Nanowires Created by FEBID/FIBID

The possibility of creating functional NWs directly by the use of the FEBID and FIBID techniques is very attractive, due to the simplicity of the fabrication processes. The NWs fabricated by these techniques usually have a complex composition and microstructure, making their physical properties very different from the pure metal molecule precursor. As described in Sect. 1.5.3 we have studied NWs using three types of gas precursors. We show now the motivation for the study of these nanosystems.

1.7.1 Pt–C NWs Created by FEBID/FIBID

These nanodeposits are one of the most commonly used in the scientific community using the FEBID/FIBID techniques [45, 46], since the gas precursor used

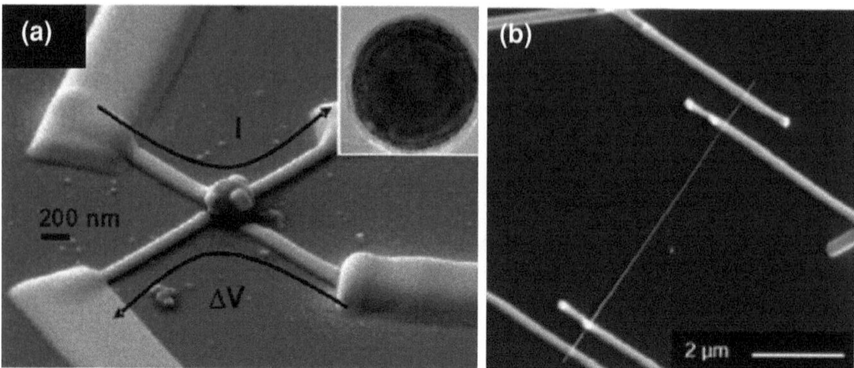

Fig. 1.18 FEBID/FIBID nanodeposits used as interconnectors for the study of **a** Fe/MgO nanospheres [88], **b** SnO$_2$ nanowires [89]. Images taken form references cited

has been commercially available in FIB or Dual Beam systems for several years. The microstructure of these deposits has been found to be inhomogeneous, with metallic clusters embedded in an amorphous C matrix. The resistivity is various orders of magnitude higher than in pure Pt using both columns (smaller in the case of using FIBID) and contradictory results have been reported regarding their electrical properties.

One of the most promising routes in nanoelectronics is the integration of nanoelements, such as carbon nanotubes, NWs, nanoparticles (NPs), organic molecules... for the creation of functional nanodevices [6, 84–89]. There exists a big challenge for developing methods to connect these nanoelements to metal electrodes. Ideal connections should have high stability, low contact resistance and ohmic behavior. One of the main applications of Pt–C nanodeposits has been the connection of NWs or NPs [88–90], due to the unique advantages FEBID/FIBID have with respect to other contacting methods. The existence of high contact resistances and Schottky barriers make a good interpretation of the devices under study difficult [90, 91]. Thus, a comprehensive understanding of the intrinsic properties of Pt–C nanodeposits is necessary for the good characterization of the nanodevices nanocontacted (Fig. 1.18).

We have made a thorough study of the microstructural, chemical and electrical properties of Pt–C NWs created by FEBID and FIBID (Chap. 5).

1.7.2 Superconducting W-Based NWs Created by FIBID

Since the discovery of superconductivity in 1911 by Kamerling Onnes [92], superconductor materials (SCs) have been one of the main research subjects in condensed-matter physics. As a result, many important findings have been made, with Fe-based SCs [93] being maybe the latest hot topic in the field.

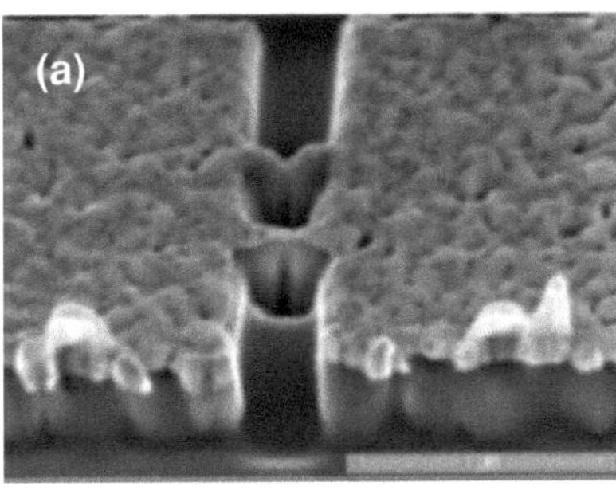

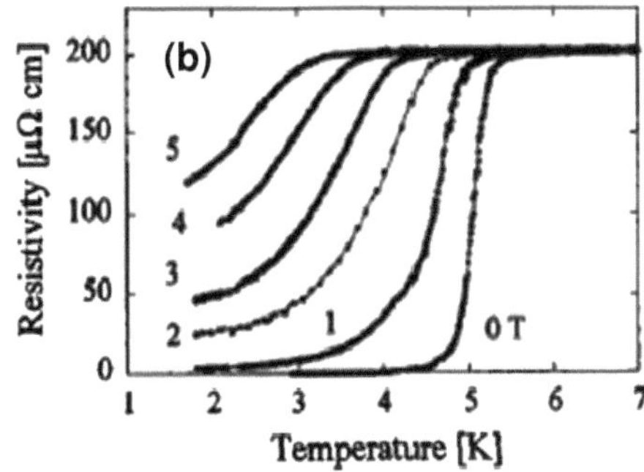

Fig. 1.19 **a** YBCO Nano-SQUID fabricated by FIB [99]. **b** Discovery by Sadki et al. that FIBID-W is superconductor above 4.2 K [55]. Images taken form references cited

As important applications of SCs, we could cite, among others, the production of high magnetic field coils, applied in Magnetic Resonance Imaging, or SQUID magnetometers. Nowadays, substantial efforts are being made in the search for better SC properties by finding new materials or by the modification of existing ones, as well as their behavior at the nanoscale. Basic studies regarding one-dimensional superconductivity [94], vortex confinement [95, 96], vortex pinning [97], etc. have been tackled recently. The fabrication of nano-SQUIDS is one of the most appealing applications of such nano-SCs [98, 99] as well as being supporting material for SC quantum bits [100, 101]. Other interesting uses are the fabrication of free-resistance SC interconnectors [102, 103], or the creation of hybrid FM-SC nanocontacts to determine the spin polarization of FM materials by Andreev reflection measurements [104]. As in other branches of nanoscience based on a top-down approximation, complex nanolithography techniques such as EBL, involving several process steps, are used to create such SC nano-structures (Fig. 1.19).

In 2004, Sadki et al. [55, 105] discovered that W-based nanodeposits created by 30 kV-FIBID, using $W(CO)_6$ as gas precursor, were SCs, with a critical temperature around 5 K. The fabrication of SC nanostructures working at liquid Helium temperatures with an easy and flexible technique such as FIBID is a very attractive option. However, before the use of FIBID-W in SC devices a good characterization of its properties should be carried out. This work is the scope of Chap. 6.

1.7.3 Magnetic Co NWs Created by FEBID

We showed in Sects. 1.2.2 and 1.4 the high impact that devices based on magnetic thin films have had in IT. The miniaturization of magnetic structures in the lateral size could result in the exploitation of new phenomena in fields such as storage and sensing. One of the most promising spintronics devices nowadays is based on the controlled guiding of domain walls (DWs) in magnetic planar NWs for the

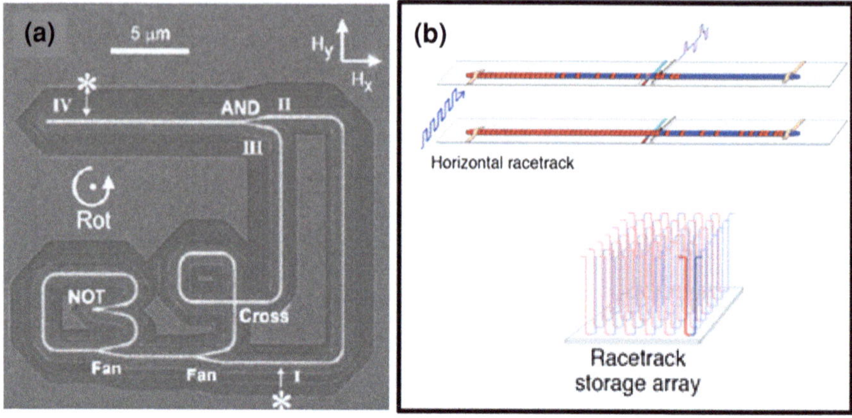

Fig. 1.20 High-impact applications of control of DWs in planar Permalloy nanowires.
a Magnetic Domain Wall Logic [106]. **b** Racetrack (*horizontal* and *vertical*) memories [107].
Images taken from references cited

transmission and storage of information. To date, there exist two main works,
using Permalloy (Fe_xNi_{1-x}) NWs:

(a) Cowburn et al. have implemented a complete magnetic logic [106], analogous
    to that currently used (CMOS technology), which relies on the movement of
    DWs in complex networks. Logical NOT, logical AND, signal fan-out and
    signal-cross-over were designed, all of them integrated in the same circuit (see
    Fig. 1.20a).
(b) Parkin et al. have proposed the so called "racetrack memories" [107]. These
    consist of non-volatile memories composed of arrays of magnetic NWs where
    the information is stored by means of tens-to-hundreds of DWs in each NW
    (see Fig. 1.20b). The working principles of these memories have been
    experimentally shown in horizontal racetracks, composed of planar NWs,
    where nanoscale current pulses move the DWs as a magnetic shift register, by
    a spin-torque effect [108].

The low metallic composition usually obtained using the FEBID technique
would seem a limitation of this nanolithography technique to fabricate highly
magnetic NWs. However, we will show that highly pure Co NWs can be created
using $Co_2(CO)_8$ as gas precursor in our system. The magnetic properties of these
NWs have been studied in Chap. 7.

1.8 Structure of the Thesis

The thesis contains 8 chapters:

In Chap. 2, the main experimental techniques used during the thesis are summarized. For fabrication: micro-(photolithography processes) and nanolithography

(Dual Beam system). And for characterization: magneto-electrical (during and after fabrication of the structures) spectroscopic (XPS and EDX), microscopy (SEM and AFM) and magnetometry (spatially-resolved MOKE) techniques, as well as the generation of high static magnetic fields.

Chapter 3 is devoted to the study of the magnetotransport properties of Fe_3O_4 epitaxial thin films, grown on MgO (001), which is a ferrimagnetic material with possible applications in spintronics. This study comprises a systematic study, as a function of temperature and film thickness, of resistivity, magnetoresistance in several geometries, anisotropic magnetoresistance, planar, ordinary and anomalous Hall effect.

The method developed for the fabrication of atomic-sized constrictions, by the control of the resistance of the device while a FIB etching process is done, is shown in Chap. 4. The study is centered on two metals: chromium and iron. The success in the use of this method in both cases, the promising MR results in the magnetic material, as well as the problems associated with the fragility of the constrictions, are presented.

Pt–C nanowires grown by both FEBID and FIBID are studied in Chap. 5. By the in situ and ex situ measurement of the electrical properties of the wires, together with a spectroscopic and microstructural characterization, a full picture for this material is obtained, where the metal–carbon composition ratio can be used to tune the electrical conduction properties.

In Chap. 6, we present the results for W-based deposits grown by FIBID. The electrical superconductor properties of this outstanding material are studied systematically for micro and nanowires. The study is completed with spectroscopy and high resolution microscopy measurements.

The compositional, magnetotransport, and magnetic measurements done in cobalt nanowires grown by FEBID is shown in Chap. 7. The influence of the beam current on the purity of the structures, the dependence of the reversal of the magnetization with shape, as well as the possibility to control domain walls in the wires are presented.

Finally, Chap. 8 sets out the main conclusions and perspectives of this thesis.

The final pages are dedicated to the bibliography of the different chapters, and also include my CV with a list of the articles published as a consequence of this work. A list of acronyms can be found at the beginning of the book.

1.9 Reproduction of Material

Some material included in this book was previously published in other scientific journal articles. I would like to acknowledge the permission obtained for the different editorials for its reproduction.

A. Fernández-Pacheco, J.M. De Teresa, J. Orna, L. Morellón, P.A. Algarabel, J.A. Pardo, and M.R. Ibarra, "Universal Scaling of the Anomalous Hall Effect in Fe3O4 thin films", Phys. Rev. B 77, 100403(R) (2008).

A. Fernández-Pacheco, J.M. De Teresa, J. Orna, L. Morellon, P.A. Algarabel, J.A. Pardo, M.R. Ibarra, C. Magen, and E. Snoeck, "Giant planar Hall effect in epitaxial Fe3O4 thin films and its temperature dependence", Phys. Rev. B 78, 212402 (2008).

A. Fernández-Pacheco, J.M. De Teresa, R. Córdoba, and M.R. Ibarra, "Conduction regimes of Pt–C nanowires grown by Focused-Ion–Beam induced deposition: Metal–insulator transition", Phys. Rev. B 79, 174204 (2009).

A. Fernández-Pacheco, J. M. De Teresa, P.A. Algarabel, J. Orna, L. Morellón, J. A. Pardo, and M.R. Ibarra, "Hall effect and magnetoresistance measurements in Fe3O4 thin films up to 30 Tesla", Appl. Phys. Lett. 95, 262108 (2009).

A. Fernández-Pacheco, J. M. De Teresa, R. Córdoba, M. R. Ibarra D. Petit, D. E. Read, L. O'Brien, E. R. Lewis, H. T. Zeng, and R.P. Cowburn, "Domain wall conduit behavior in cobalt nanowires grown by Focused-Electron-Beam Induced Deposition", Appl. Phys. Lett. 94, 192509 (2009).

I. Guillamón, H. Suderow, S. Vieira, A. Fernández-Pacheco, J. Sesé, R. Córdoba, J.M. De Teresa, and M.R. Ibarra, "Nanoscale superconducting properties of amorphous W-based deposits grown with a focused-ion-beam", New J. Phys. 10, 093005 (2008).

A. Fernández-Pacheco, J. M. De Teresa, R. Córdoba, and M. R. Ibarra, "Magnetotransport properties of high-quality cobalt nanowires grown by focused-electron-beam-induced deposition", J.Phys D: Appl. Phys. 42, 055005 (2009).

A. Fernández-Pacheco, J. M. De Teresa, R. Córdoba, M. R. Ibarra D. Petit, D. E. Read, L. O'Brien, E. R. Lewis, H. T. Zeng, and R.P. Cowburn, "Systematic study of the magnetization reversal in Co wires grown by focused-electron-beam-induced deposition", Nanotechnology 20, 475704 (2009).

A. Fernández-Pacheco, J. M. De Teresa, R. Córdoba, and M.R. Ibarra, "Exploring the conduction in atomic-sized metallic constrictions created by controlled ion etching", Nanotechnology 19, 415302 (2008).

A. Fernández-Pacheco, J.M. De Teresa, R. Córdoba, and M.R. Ibarra, "Tunneling and anisotropic-tunneling magnetoresistance in iron nanoconstrictions fabricated by focused-ion-beam", MRS proc. DD02-03, 1181 (2009).

J. M. De Teresa, A. Fernández-Pacheco, R. Córdoba, J. Sesé, M. R. Ibarra, I. Guillamón, H. Suderow, and S. Vieira, "Transport properties of superconducting amorphous W-based nanowires fabricated by focused-ion-beam-induced-deposition for applications in Nanotechnology", MRS proc. CC04-09, 1880 (2009).

J.M. De Teresa, R. Córdoba, A. Fernández-Pacheco, O. Montero, P. Strichovanec, and M.R. Ibarra, "Origin of the Difference in the Resistivity of As-Grown Focused-Ion and Focused-Electron-Beam-Induced Pt Nanodeposits", J. Nanomat. 2009, 936863 (2009).

I. Guillamón, H. Suderow, A. Fernández-Pacheco, J. Sesé, R. Córdoba, J. M. De Teresa, M.R. Ibarra, and S. Vieira, "Direct observation of melting in a 2-D superconducting vortex lattice" Nat. Phys. 5, 651 (2009).

References

1. R.P. Feynman, There's Plenty of Room at the Bottom, Engineering and Science. **23**, 22 (1960)
2. C.M. Vest, Dissertation, Massachusetts Institute of Technology, 2005
3. S. Bensasson, Technology Roadmap for Nanoelectronics In: Ramón Compañó, (ed.) Directorate-General Information Society European Commission (2000)
4. G. Moore, Cramming more components onto integrated circuits. Electronics **38**, 114 (1965)
5. http://www.greentechmedia.com/content/images/articles/varian-moores-law-graph.jpg
6. J. Ferrer, V.M. García-Suárez, From microelectronics to molecular spintronics: an explorer's travelling guide. J. Mater. Chem. **19**, 1696 (2009)
7. L.R. Harriott, Limits of lithography. IEEE Proc. **89**, 366 (2001)
8. T. Ito, S. Okazaky, Pushing the limits of lithography. Nature **406**, 1027 (2000)
9. R. Chau, B. Doyle, S. Datta, J. Kavalieros, K. Zhang, Integrated nanoelectronics for the future. Nat. Mater. **6**, 810 (2007)
10. A. Correira, P.A. Serena, *Nanociencia y Nanotecnología en España* (Fundación Phantoms, 2008), p. 7
11. É. du Trémolet de Lacheisserie, D. Gignoux, M. Schlenker, *Magnetism*, vol. 2 ("Materials and applications"), Chap. 21 (Springer, 2005), p. 305
12. MRS Bulletin, *Materials for Magnetic Data Storage*, vol. 31 (May 2006)
13. E.E. Fullerton, *Materials Requirements for Next-Generation Magnetic Storage and Memory Applications* (SNOW School Lecture, Nancy, France, 2008)
14. P. Grünberg, R. Schreiber, Y. Pang, M.B. Brodsky, H. Sowers, Layered magnetic structures: evidence for antiferromagnetic coupling of Fe layers across Cr interlayers. Phys. Rev. Lett. **57**, 2442 (1986)
15. M.T. Béal-Monod, Ruderman–Kittel–Kasuya–Yosida indirect interaction in two dimensions. Phys. Rev. B **36**, 8835 (1987)
16. N.F. Mott, Electrons in transition metals. Adv. Phys. **13**, 325 (1964)
17. A. Fert, I.A. Campbell, Two-current conduction in nickel. Phys. Rev. Lett. **21**, 1190 (1968)
18. C. Chappert, A. Fert, F. Nguyen Van Dau, The emergence of spin electronics in data storage. Nat. Mater. **6**, 813 (2007)
19. R. Jansen, The spin-valve transistor: a review and outlook. J. Phys. D Appl. Phys. **36**, R289 (2003)
20. I. Zutic, J. Fabian, S. Dasma, Spintronics: fundamentals and applications. Rev. Mod. Phys. **76**, 323 (2004)
21. E.E. Fullerton, I.K. Schüller, The 2007 Nobel Prize in physics: magnetism and transport at the nanoscale. ACS Nano **5**, 384 (2007)
22. A. Fert, Nobel lecture: origin, development, and future of spintronics. Rev. Mod. Phys. **80**, 1517 (2008)
23. P. Grünberg, Nobel lecture: from spin waves to giant magnetoresistance and beyond. Rev. Mod. Phys. **80**, 1531 (2008)
24. M.N. Baibich, J.M. Broto, A. Fert, F. Nguyen Van Dau, F. Petroff, P. Etienne, G. Creuzet, A. Friederich, J. Chazelas, Giant magnetoresistance of (001)Fe/(001)Cr magnetic superlattices. Phys. Rev. Lett. **61**, 2472 (1988)
25. G.P. Binasch, F. Grünberg, F. Saurenbach, W. Zinn, Enhanced magnetoresistance in layered magnetic structures with antiferromagnetic interlayer exchange. Phys. Rev. B **39**, 4828 (1989)
26. A. Yanase, K. Siratori, Band structure in the high temperature phase of Fe_3O_4. J. Phys. Soc. Jpn. **53**, 312 (1984)
27. Z. Zhang, S. Satpathy, Electron states, magnetism, and the Verwey transition in magnetite. Phys. Rev. B **44**, 13319 (1991)
28. S.F. Alvarado, W. Eib, F. Meier, D.T. Pierce, K. Sattler, H.C. Siegmann, Observation of spin-polarized electron levels in ferrites. Phys. Rev. Lett. **34**, 319 (1975)

29. M. Fonin, Yu.S. Dedkov, R. Pentcheva, U. Rüdiger, G. Güntherodt, Magnetite: a search for the half-metallic state. J. Phys. Condens. Matter **19**, 315217 (2007)
30. G. Hu, Y. Suzuki, Negative spin polarization of Fe_3O_4 in magnetite/manganite-based junctions. Phys. Rev. Lett. **89**, 276601 (2002)
31. J.Y.T. Wei, N.-C. Yeh, R.P. Vasquez, A. Gupta, Tunneling evidence of half-metallicity in epitaxial films of ferromagnetic perovskite manganites and ferrimagnetic magnetite. J. Appl. Phys. **83**, 7366 (1998)
32. W.E. Picket, J.S. Moodera, Half metallic magnets. Phys. Today **54**(5), 39 (2001)
33. C.M. Fang, G.A. de Wijs, R.A. de Groot, Spin-polarization in Half-metals. J. Appl. Phys. **91**, 8340 (2002)
34. M. Jullière, Tunneling between ferromagnetic films. Phys. Lett. **54A**, 225 (1975)
35. V. Cros, *From the Origin to Recent Developments in Spintronics* (Snow School Lecture, Nancy, France, 2008)
36. J.M. De Teresa, A. Barthélèmy, A. Fert, J.P. Contour, R. Lyonnet, F. Montaigne, P. Seneor, Role of metal-oxide interface in determining the spin polarization of magnetic tunnel junctions. Science **286**, 507 (1999)
37. W.H. Butler, X.G. Zhang, T.C. Schulthess, J.M. MacLaren, Spin-dependent tunneling conductance of Fe|MgO|Fe sandwiches. Phys. Rev. B **63**, 054416 (2001)
38. S. Yuasa, D.D. Djayaprawira, Giant tunnel magnetoresistance in magnetic tunnel junctions with a crystalline MgO(001) barrier. J. Phys. D Appl. Phys. **40**, R337 (2007)
39. J. Moodera, L.R. Kinder, R.M. Wong, R. Meservey, Large magnetoresistance at room temperature in ferromagnetic thin film tunnel junctions. Phys. Rev. Lett. **74**, 3273 (1995)
40. M. Bowen, M. Bibes, A. Barthélémy, J.P. Contour, A. Anane, Y. Lemaître, A. Fert, Nearly total spin polarization in $La_{2/3}Sr_{1/3}MnO_3$ from tunneling experiments. Appl. Phys. Lett. **82**, 233 (2003)
41. S. Mao, Y. Chen, F. Liu, X. Chen, B. Xu, P. Lu, M. Patwari, H. Xi, C. Chang, B. Miller, D. Menard, B. Pant, J. Loven, K. Duxstad, S. Li, Z. Zhang, A. Johnston, R. Lamberton, M. Gubbins, T. McLaughlin, J. Gadbois, J. Ding, B. Cross, S. Xue, P. Ryan, Commercial TMR heads for hard disk drives: characterization and extendibility at 300 Gbit/in.^{2}. IEEE Trans. Magn. **42**, 97 (2006)
42. L. Giannuzzi, F. Stevie, *Introduction to Focused Ion Beams: Instrumentation, Techniques, Theory and Practices* (Springer Science + Business Media, Boston, 2005)
43. J. Orloff, M. Utlaut, L. Swanson, *High Resolution Focused Ion Beams: FIB and Its Applications* (Kluwer Academic/Plenum Publishers, NY, 2003)
44. N. Silvis-Clividjan, C.W. Hagen, P. Kruit, Spatial resolution limits in electron-beam-induced deposition. J. Appl. Phys. **98**, 084905 (2005)
45. I. Utke, P. Hoffmann, J. Melngailis, Gas-assisted focused electron beam and ion beam processing and fabrication. J. Vac. Sci. Technol. B **26**, 1197 (2008)
46. W.F. van Dorp, C.W. Hagen, A critical literature review of focused electron beam induced deposition. J. Appl. Phys. **104**, 081301 (2008)
47. A. Botman, J.J.L. Mulders, R. Weemaes, S. Mentik, Purification of platinum and gold structures after electron-beam-induced deposition. Nanotechnology **17**, 3779 (2006)
48. R.M. Langford, T.-X. Wang, D. Ozkaya, Reducing the resistivity of electron and ion beam assisted deposited Pt. Microelectron. Eng. **84**, 748 (2007)
49. S. Matsui, Y. Ochiai, Focused ion beam applications to solid state devices. Nanotechnology **7**, 247 (1996)
50. W. Li, P.A. Warburton, Low-current focused-ion-beam induced deposition of three-dimensional tungsten nanoscale conductors. Nanotechnology **18**, 485305 (2007)
51. A. De Marco, J. Melngailis, Maskless fabrication of JFETs via. Focused ion beams. Solid State Electron. **48**, 481833 (2004)
52. G.C. Gazzadi, S. Frabboni, Fabrication of 5 nm gap pillar electrodes by electron-beam Pt deposition. J. Vac. Sci. Technol. B **23**, L1 (2005)
53. A. Romano-Rodríguez, F. Hernández-Ramirez, Dual-beam focused ion beam (FIB): a prototyping tool for micro and nanofabrication. Microelectron. Eng. **84**, 789 (2007)

54. I. Utke, P. Hoffmann, R. Berger, L. Scandella, High resolution magnetic force microscopy supertips produced by focused electron beam induced deposition. Appl. Phys. Lett. **80**, 4792 (2002)

55. E.S. Sadki, S. Ooi, K. Hirata, Focused-ion-beam-induced deposition of superconducting nanowires. Appl. Phys. Lett. **85**, 6206 (2004)

56. N. Agraït, A.L. Yeyati, J.M. Van Ruitenbeek, Quantum properties of atomic-sized conductors. Phys. Rep. **377**, 81 (2003)

57. Yu.V. Sharvin, A possible method for studying Fermi surfaces. Sov. Phys.-JETP **21**, 655 (1965)

58. R. Landauer, Spatial variation of currents and fields due to localized scatterers in metallic conduction. IBM J. Res. Dev. **1**, 223 (1957)

59. E. Scheer, N. Agra, J.C. Cuevas, A.L. Yeyati, B. Ludoph, A. Martín-Rodero, G.R. Bollinger, J.M. van Ruitenbeek, C. Urbina, The signature of chemical valence in the electrical conduction through a single-atom contact. Nature **394**, 154 (1998)

60. J.M. van Ruitenbeek, A. Alvarez, I. Pineyro, C. Grahmann, P. Joyez, M.H. Devoret, D. Esteve, C. Urbina, Adjustable nanofabricated atomic size contacts. Rev. Sci. Instrum. **67**, 108 (1996)

61. R. Wiesendanger, *Scanning Probe Microscopy and Spectroscopy* (Cambridge University Press, Cambridge, 1994)

62. H. Park, A.K.L. Lim, A.P. Alivisatos, J. Park, P.L. Mceuen, Fabrication of metallic electrodes with nanometer separation by electromigration. Appl. Phys. Lett. **75**, 301 (1999)

63. A.F. Morpurgo, C.M. Marcus, D.B. Robinson, Controlled fabrication of metallic electrodes with atomic separation. Appl. Phys. Lett. **74**, 2084 (1999)

64. K.M. Schep, P.J. Kelly, G.E.W. Bauer, Giant magnetoresistance without defect scattering. Phys. Rev. Lett. **74**, 586 (1995)

65. K. Wellock, S.J.C.H. Theeuwen, J. Caro, N.N. Gribov, R.P. Van Gorkom, S. Radelaar, F.D. Tichelaar, B.J. Hickey, C.H. Marrows, Giant magnetoresistance of magnetic multilayer point contacts. Phys. Rev. B **60**, 10291 (1999)

66. N. García, M. Muñoz, Y.W. Zhao, Magnetoresistance in excess of 200% in ballistic Ni nanocontacts at room temperature and 100 Oe. Phys. Rev. Lett. **82**, 2923 (1999)

67. N. García, M. Muñoz, G.G. Qian, H. Rohrer, I.G. Saveliev, Y.W. Zhao, Ballistic magnetoresistance in a magnetic nanometer sized contact: an effective gate for spintronics. Appl. Phys. Lett. **79**, 4550 (2001)

68. J.J. Versluijs, M.A. Bari, J.M.D. Coey, Magnetoresistance of half-metallic oxide nanocontacts. Phys. Rev. Lett. **87**, 026601 (2001)

69. S.H. Chung, M. Muñoz, N. García, W.F. Egelhoff, R.D. Gomez, Universal scaling of ballistic magnetoresistance in magnetic nanocontacts. Phys. Rev. Lett. **89**, 287203 (2002)

70. P. Bruno, Geometrically constrained magnetic wall. Phys. Rev. Lett. **83**, 2425 (1999)

71. W.F. Egelhoff Jr., L. Gan, H. Ettedgui, Y. Kadmon, C.J. Powell, P.J. Chen, A.J. Shapiro, R.D. McMichael, J.J. Mallett, T.P. Moffat, M.D. Stiles, E.B. Svedberg, Artifacts in ballistic magnetoresistance measurements (invited). J. Appl. Phys. **95**, 7554 (2004)

72. M.I. Montero, R.K. Dumas, G. Liu, M. Viret, O.M. Stoll, W.A.A. Macedo, I.K. Schuller, Magnetoresistance of mechanically stable Co nanoconstrictions. Phys. Rev. B **70**, 184418 (2004)

73. O. Ozatay, P. Chalsani, N.C. Emley, I.N. Krivorotov, R.A. Buhrman, Magnetoresistance and magnetostriction effects in ballistic ferromagnetic nanoconstrictions. J. Appl. Phys. **95**, 7315 (2004)

74. B. Doudin, M. Viret, Ballistic magnetoresistance? J. Phys. Condens. Matter **20**, 083201 (2008)

75. J. Smit, Magnetoresistance of ferromagnetic metals and alloys at low temperatures. Physica **17**, 612 (1951)

76. R. Potter, Magnetoresistance anisotropy in ferromagnetic NiCu alloys. Phys. Rev. B **10**, 4626 (1974)

77. P. Gambardella, A. Dallmeyer, K. Maiti, M.C. Malagoli, W. Eberhardt, K. Kern, C. Carbone, Ferromagnetism in one-dimensional monatomic metal chains. Nature **416**, 301 (2002)
78. J. Velev, R.F. Sabirianov, S.S. Jaswal, E.Y. Tsymbal, Ballistic anisotropic magnetoresistance. Phys. Rev. Lett. **94**, 127203 (2005)
79. D. Jacob, J. Fernández-Rossier, J.J. Palacios, Anisotropic magnetoresistance in nanocontacts. Phys. Rev. B **77**, 165412 (2008)
80. A. Sokolov, C. Zhang, E.Y. Tsymbal, J. Redepenning, B. Doudin, Quantized magnetoresistance in atomic-size contacts. Nat. Nanotechnol. **2**, 171 (2007)
81. M. Viret, M. Gauberac, F. Ott, C. Fermon, C. Barreteau, G. Autes, R. Guirado López, Giant anisotropic magneto-resistance in ferromagnetic atomic contacts. Eur. Phys. J. B **51**, 1 (2006)
82. S.-F. Shi, D.C. Ralph, Atomic motion in ferromagnetic break junctions. Nat. Nanotechnol. **2**, 522 (2007)
83. M. Häfner, J.K. Viljas, J.C. Cuevas, Theory of anisotropic magnetoresistance in atomic-sized ferromagnetic metal contacts. Phys. Rev. B **79**, 140410(R) (2009)
84. M.P. Anantram, F.L. Leonard, Physics of carbon nanotube electronic devices. Rep. Prog. Phys. **69**, 507 (2006)
85. L.E. Hueso, J.M. Pruneda, V. Ferrari, G. Burnell, J.P. Valdés-Herrera, B.D. Simons, P.B. Littlewood, E. Artacho, A. Fert, N.D. Mathur, Transformation of spin information into large electrical signals using carbon nanotubes. Nature **445**, 410 (2007)
86. Z.-M. Liao, Y.-D. Li, J. Xu, J.-M. Zhang, K. Xia, D.-P. Yu, Spin-filter effect in magnetite nanowire. Nano Lett. **6**, 1087 (2006)
87. R.H.M. Smit, Y. Noat, C. Untiedt, N.D. Lang, M.C. van Hemert, J.M. van Ruitenbeek, Measurement of the conductance of a hydrogen molecule. Nature **419**, 906 (2002)
88. C. Martínez-Boubeta, Ll. Balcells, C. Monty, P. Ordejon, B. Martínez, Tunneling spectroscopy in core/shell structured Fe/MgO nanospheres. Appl. Phys. Lett. **94**, 062507 (2009)
89. F. Hernández-Ramírez, S. Barth, A. Tarancón, O. Casals, E. Pellicer, J. Rodríguez, A. Romano-Rodríguez, J.R. Morante, S. Mathur, Water vapor detection with individual tin oxide nanowires. Nanotechnology **18**, 424016 (2007)
90. F. Hernández-Ramírez, A. Tarancón, O. Casals, E. Pellicer, J. Rodríguez, A. Romano-Rodríguez, J.R. Morante, S. Barth, S. Mathur, Electrical properties of individual tin oxide nanowires contacted to platinum electrodes. Phys. Rev. B **76**, 085429 (2007)
91. T. Scwamb, B.R. Burg, N.C. Schirmer, D. Poulikakos, On the effect of the electrical contact resistance in nanodevices. Appl. Phys. Lett. **92**, 243106 (2008)
92. H. Kamerlingh Onnes, The resistance of pure mercury at helium temperatures. Leiden Commun. **120b** (1911)
93. Y. Kamihara, T. Watanabe, M. Hirano, H. Hosono, Iron-Based Layered Superconductor La[O$_{1-x}$F$_x$]FeAs (x = 0.05−0.12) with T$_c$ = 26 K. J. Am. Chem. Soc. **130**, 3296 (2008)
94. K.Yu. Arutyunov, D.S. Golubev, A.D. Zaikin, Superconductivity in one dimension. Phys. Rep. **464**, 1 (2008)
95. L.F. Chibotaru, A. Ceulemans, V. Bruyndoncx, V.V. Moschalkov, Symmetry-induced formation of antivortices in mesoscopic superconductors. Nature **408**, 833 (2000)
96. I.V. Grigorieva, A.K. Geim, S.V. Dubonos, K.S. Novoselov, Long-range nonlocal flow of vortices in narrow superconducting channels. Phys. Rev. Lett. **92**, 237001 (2004)
97. M. Velez, J.I. Martín, J.E. Villegas, A. Hoffmann, E.M. González, J.L. Vicent, I.K. Schuller, Superconducting vortex pinning with artificial magnetic nanostructures. J. Magn. Magn. Mater. **320**, 2547 (2008)
98. C. Granata, E. Exposito, A. Vettoliere, L. Petti, M. Russo, An integrated superconductive magnetic nanosensor for high-sensitivity nanoscale applications. Nanotechnology **19**, 275501 (2008)
99. C.H. Wu, Y.T. Chou, W. CKuo, J.H. Chen, L.M. Wang, J.C. Chen, K.L. Chen, U.C. Sou, H.C. Yang, J.T. Jeng, Fabrication and characterization of high-Tc YBa$_2$Cu$_3$O$_{7-x}$ nanoSQUIDs made by focused ion beam milling. Nanotechnology **19**, 315304 (2008)

100. J. Clarke, F.K. Wilhelm, Superconducting quantum bits. Nature **453**, 1031 (2008)
101. A.J. Legget, Superconducting qubits—a major roadblock dissolved? Science **296**, 861 (2002)
102. A.Y. Kasumov, K. Tsukagoshi, M. Kawamura, T. Kobayashi, Y. Aoyagi, K. Senba, T. Kodama, H. Nishikawa, I. Ikemoto, K. Kikuchi, V.T. Volkov, A.Yu. Kasumov, R. Deblock, S. Guéron, H. Bouchiat, Proximity effect in a superconductor-metallofullerene-superconductor molecular junction. Phys. Rev. B **72**, 033414 (2005)
103. A. Shailos, W. Nativel, A. Kasumov, C. Collet, M. Ferrier, S. Guéron, R. Deblock, H. Bouchiat, Proximity effect and multiple Andreev reflections in few-layer graphene. Europhys. Lett. **79**, 57008 (2007)
104. R.J. Soulen Jr., J.M. Byers, M.S. Osofsky, B. Nadgorny, T. Ambrose, S.F. Cheng, P.R. Broussard, C.T. Tanaka, J. Nowak, J.S. Moodera, A. Barry, J.M.D. Coey, Measuring the spin polarization of a metal with a superconducting point contact. Science **282**, 85 (1998)
105. E.S. Sadki, S. Ooi, K. Hirata, Focused-ion-beam-induced deposition of superconducting thin films. Physica C **426–431**, 1547 (2005)
106. D.A. Alwood, G. Xiong, C.C. Faulkner, D. Atkinson, D. Petit, R.P. Cowburn, Magnetic domain-wall logic. Science **309**, 1688 (2005)
107. S.S.P. Parkin, M. Hayashi, L. Thomas, Magnetic domain-wall racetrack memory. Science **320**, 190 (2008)
108. D.C. Ralph, M.D. Stiles, Spin transfer torques. J. Magn. Magn. Mater. **320**, 1190 (2008)

Chapter 2
Experimental Techniques

In this chapter we report on the experimental techniques used during this thesis both in the sample patterning and characterization. An important part will be focused on the lithography techniques and processes (micro and nano) used for the creation of structures with a well defined geometry in these scales. The setup of an experimental installation for the study of magnetotransport properties as a function of temperature, as well as that for electrical measurements inside a Dual Beam chamber, will be also explained. Spectroscopy (EDX and XPS), magnetometry (MOKE) and microscopy (SEM and AFM) techniques used during the work are included in this chapter. Most of the instruments described are located at the Institute of Nanoscience of Aragón (INA) and at the Institute of Science Materials of Aragón (ICMA). Spatially resolved MOKE and AFM were done over a two-month stage with the research group of Prof. Cowburn at Imperial College, London. Magnetotransport experiments at high static fields were performed at the High Field Magnet Laboratory of the University of Nijmegen, The Netherlands.

2.1 Lithography Techniques

Lithography is the group of techniques used to transfer previously determined patterns on a substrate. The typical size of these structures is micro- and nanometric. Depending on the type of lithography technique used, the transfer is performed in a different way, which defines the resolution that can be attained. In this thesis, we have used two large lithography groups, both of them within the top-down approximation:

1. Optical lithography: these techniques have been used in a clean room environment. This is by far the most common lithography technique in microelectronic fabrication. It permits the fabrication of micrometric structures.

A. Fernandez-Pacheco, *Studies of Nanoconstrictions, Nanowires and Fe₃O₄ Thin Films*, Springer Theses, DOI: 10.1007/978-3-642-15801-8_2,

2. Lithography using focused electron and ion beams (Dual Beam system): with these techniques, nanostructures below 100 nm can be patterned.

We should remark that the equipment used was installed in the premises of the Institute of Nanoscience of Aragón (INA) in Zaragoza, during this thesis. Thus, the procedures for the majority of the processes that will be shown here were carried out by myself, together with the technical staff of the Lithography Laboratory (R. Valero, I. Rivas, and R. Córdoba), and Doctor Javier Sesé.

2.1.1 Optical Lithography

All the photolithography processes have been performed in a 10,000 class-clean room, at the INA. Clean rooms are special rooms with well-controlled ambient conditions. Temperature, humidity, differential pressure and flux of air, as well as lightning and electrostatic protection have to fulfill determined standards. Class = 10,000 means that the number of particles of size ≤ 0.5 μm per cubic meter should not be higher than 10^4. A deviation of these parameters can alter the processes, as well as modify their quality.

We will give a short overview of the equipment and processes used by explaining one particular example patterned in this thesis: lithography in Fe_3O_4 thin films, whose results will be presented in Chap. 3.

The starting point is to create a computer layout for the specific application. We have used the program Clewin® for this. In Fig. 2.1. we show the mask design, where the different colors symbolize different layers. Since we wish to have two materials on top of the substrate (in this case MgO) with different shapes, it will be necessary to perform two lithographic processes. For both patterns to be well oriented one with respect to the other, lithography marks are added to the design (see Fig. 2.1c).

Once the pattern is done, two photolithography masks (one for each step) are made. These masks consist of a glass plate having the desired pattern in the form of a thin (~ 100 nm) chromium layer. The masks were manufactured by Delta Mask®, company located in the Netherlands.

As shown in Fig. 2.2, the lithography process involves the following steps:

First step: The Fe_3O_4 film on top of MgO is patterned with the shape shown in Fig. 2.1b—left. The processes involved are:

a. A thin film of Fe_3O_4 is epitaxially deposited on MgO (001) by Pulsed Laser Deposition. This work is part of the thesis of Orna [1].
b. The Fe_3O_4 thin film is spin coated with a photoresist. This is a polymeric photosensitive material that can be spun onto the wafer in liquid form. The spinning speed and photoresist viscosity will determine the final resist thickness. For the first step, we used a typical thickness of 2.4 μm.
c. The substrate is soft-baked during 50 s at 110°C on a hot plate, in order to remove the solvents from the resist, minimize the stress, and improve the adhesion.

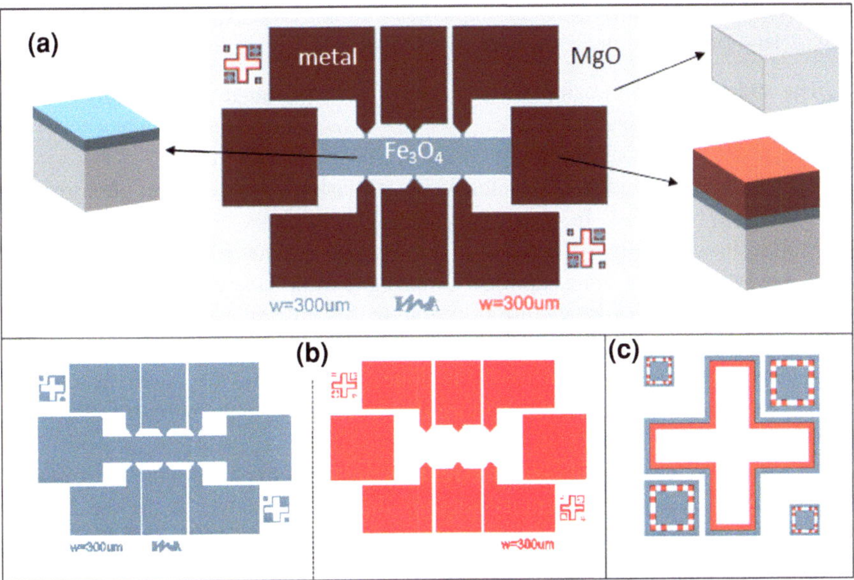

Fig. 2.1 Lithography mask designed by computer. **a** Top and 3D view of the mask. The *red, blue* and *grey* symbols symbolize a metallic material, Fe_3O_4 and the substrate MgO, respectively. **b** The two patterns included in the design. Each pattern requires a separated complete lithography process. **c** Marks used, which are necessary for the good alignment of the two steps

d. The mask is aligned to the wafer using a *Karl Suss*® mask aligner, and the photoresist is exposed to a UV source, passing through the photomask (Hg light source, h-line: $\lambda = 405$ nm) during 7 s. The areas of the mask with chromium are completely opaque to the light and complementary areas are transparent. The light passes through the mask. The zones of photoresist which are exposed to light undergo a chemical reaction upon this exposure. As the resist used is positive, exposed zones become more soluble in the developer used.

e. The resist is developed, leaving regions that are covered by photoresist and complementary regions that are not covered (as the resist is positive, it takes the shape of the structure designed, the blue structure in Fig. 2.1b—left).

f. The remaining resist is hard baked (125°C, 2 min on a hot plate) to harden it for next processes.

g. The sample is introduced into a *Sistec*® ion etching equipment, where ~ 300 eV argon ions impinge on the sample, etching it. As the Fe_3O_4 films have a maximum thickness of 40 nm, unprotected parts are removed, whereas those which have resist on top remain unharmed by this process. By the use of an electron neutralizer gun, the ions are neutralized ($Ar^+ \rightarrow Ar^0$) before arriving at the sample, avoiding charging effects at the surface of the insulator MgO substrate, which would eventually spoil the etching process after a certain point.

h. The remaining resist is removed by acetone immersion.

Second step: Metallic pads are deposited on top of the Fe_3O_4, with the shape of Fig. 2.1b—right. The processes are as follows:

a. The sample is spin coated with an image-reversal (negative) photoresist. The thickness chosen was of 3 μm.
b. The resist is soft-baked during 2 min at 100°C on a hot plate.
c. The mask is aligned to the wafer, using a mask aligner, and the photoresist is exposed (t = 10 s). In principle, the exposed zones become more soluble. Marks shown in Fig. 2.1c are used for a good alignment of both structures.
d. A reversal bake is done, at 130 °C during 2 min. This process crosslinks the exposed photoresist, becoming insoluble in the developer. The unexposed areas still behave like a normal unexposed positive resist.
e. A flood exposure, without mask, is done (27 s). Thus, the zones which were first exposed (step c) are much more insoluble in comparison with those which were not.
f. The resist is developed, leaving regions in the sample that are covered by photoresist and complementary regions that are not covered (as the resist is negative, resist forms the complementary structure of that designed, Fig. 2.2b—right).
g. The sample is introduced into an *Edwards*® electron-gun evaporator. A 100 nm thick metal layer (typically polycrystalline gold or aluminum) is evaporated. Usually a thin layer (around 10 nm) of chromium is deposited beforehand, for good adhesion of the metallic layer.
h. The sample is immersed in acetone. The profile of the negative resist, in contrast to the positive one (see in detail the resist profiles in Fig. 2.2), favors the complete removal of those parts of metal with resist behind them. This type of process is called "lift-off".

An optical image of the sample, with gold as metallic layer, is shown in Fig. 2.3a. We should remark that MgO substrates become cracked when immersed in an ultrasound bath. This makes this lithography process a bit more complicated than when using, for example, normal silicon substrates.

The pads are finally connected by ultrasonic wire bonding to a chip carrier, with a *Kulicke & Soffa Ltd*® instrument. The chip carrier (Fig. 2.3b) has macroscopic connections, which can be manipulated in a conventional way.

The lift-off process explained has been used in other types of designs for applications using the Dual Beam system (Sect. 2.1.2). In this case, metallic circuits are patterned by photolithography, consequently nano-lithographied by the Dual Beam equipment. The substrate chosen for these applications is a silicon wafer, with a 150–250 nm thick insulator on top of it (either Si_3N_4 deposited by Plasma-Enhanced-Chemical-Vapor-Deposition, using *Sistec*® PECVD equipment, or thermally oxidized SiO_2). This layer insulates electrically the metallic pads and, at the same time, electrons or ions can be partially evacuated thorough the Si substrate, avoiding charging effects which ruin the nanolithography process.

An image of the equipment used for the photolithography is shown in Fig. 2.4. The minimum line width obtained routinely in our laboratory in these processes is

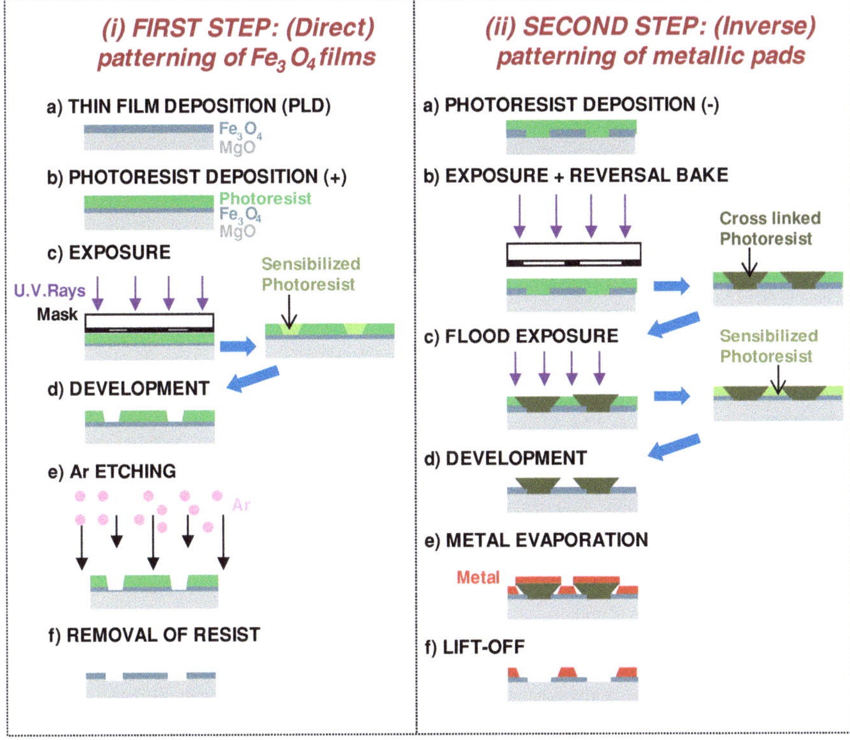

Fig. 2.2 Photolithography of Fe_3O_4 thin films. The two layers are patterned in two separate processes. In this scheme the baking steps, explained in the text, are omitted. The photoresist profiles are exaggerated for pedagogical reasons

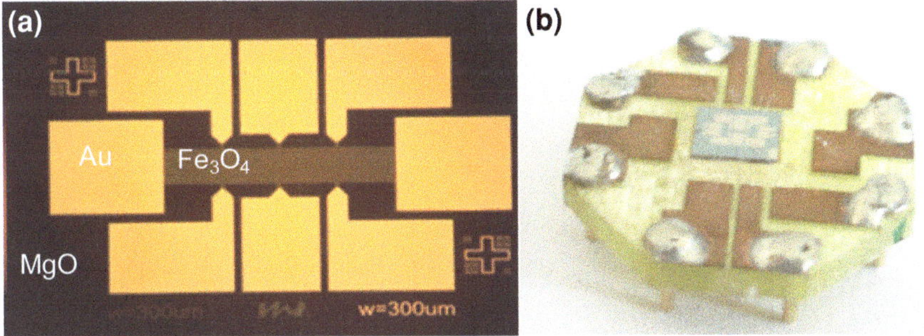

Fig. 2.3   a Optical image of one Fe_3O_4 bar with gold pads on top, the result of a two-step lithography process. **b** Sample mounted onto a chip carrier and microcontacted. The chip carrier was designed together with the Dr. Óscar Montero and the technician Carlos Martín

around 3–4 μm (better in the case of positive resist). The final resolution in an optical process is given by the diffraction limit ($\lambda/2$, λ being the light wavelength), although other effects such as optical aberrations of the system, the exposure mode

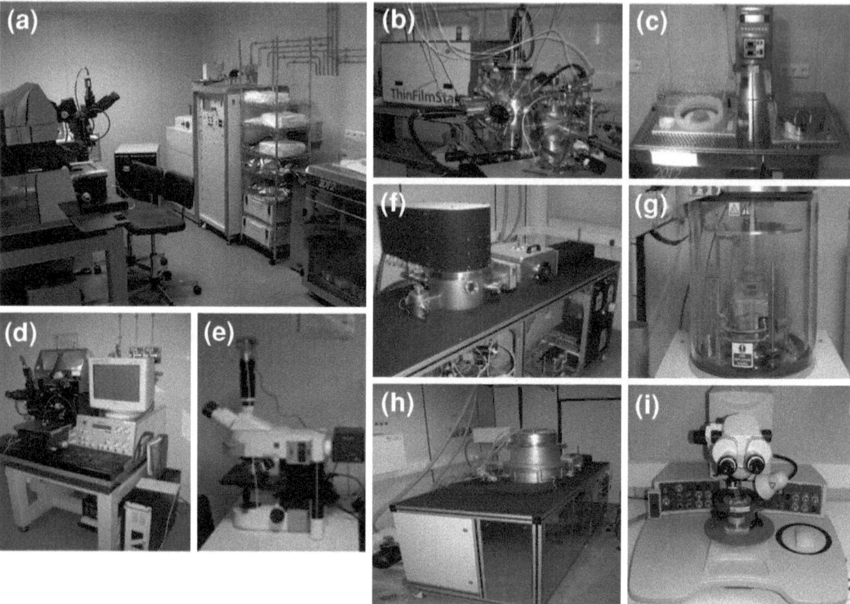

Fig. 2.4 Images of the instruments used for photolithography. **a** General view of the clean room. **b** Pulsed-Laser-Deposition equipment. **c** Spin coater and hot plate. **d** Mask aligner. **e** Optical microscope. **f** Plasma-Enhanced-Chemical–Vapor-Deposition equipment. **g** Electron-gun evaporator. **h** Ion milling equipment. **i** Wire bonding equipment

and the chemical process linked to the resistance are what actually dictate the limits. In the semiconductor industry, critical dimensions below 100 nm are nowadays routinely patterned in integrated circuits [2, 3]. Huge steppers working in projection mode, KrF or ArF lasers and immersion lenses to increase the numerical aperture are used to push the limits of this technology [2, 3]. For a more detailed explanation of these processes, as well as of general aspects of lithography, see Ref. [4, 5].

2.1.2 Dual Beam System

A wide description of a Dual Beam, integrating a SEM and a FIB column for nanolithography, was given in the introductory chapter, due to its relevance on the thesis and novel character of the processes performed. In general, some of the main applications of these systems are [6]:

1. FIB etching and simultaneous SEM imaging.
2. Imaging with both beams: although the electron beam is the primary imaging tool in a dual beam, the imaging from the ion and electron beams is often complementary. The collection of secondary ions is possible when using FIB.

3. Cross-section fabrication and analysis.
4. Local deposition of materials by introducing precursor gases inside the chamber (FEBID and FIBID, Sect. 1.5.3).
5. Lamella preparation for transmission electron microscopy (TEM) analysis. Thin specimens, transparent to high-energy electrons can be fabricated. Before the etching, a protective layer is deposited. After several progressive thinning processes, a micromanipulator takes the lamella to the TEM grid. This method is becoming popular in microscopy laboratories as an alternative method to the manual methods.

In Fig. 2.5 images of the *Nova Nanolab 200* by FEI® experimental system are shown, including columns, detectors, manipulator, and needles for gas injection (GIS). The processes are carried out in a chamber at high vacuum (turbo-molecular pump), whereas both columns need ultra-high vacuum (ionic pumps).

As commented in Sect. 1.5 this system has been used to create nanostructures:

- By simultaneous FIB etching and electrical resistance control, we have created atomic-sized nanoconstrictions, using the SEM column for imaging the process.
- By the injection of a gas precursor using the GIS, we have deposited nanowires of different materials using both focused columns (FEBID and FIBID).

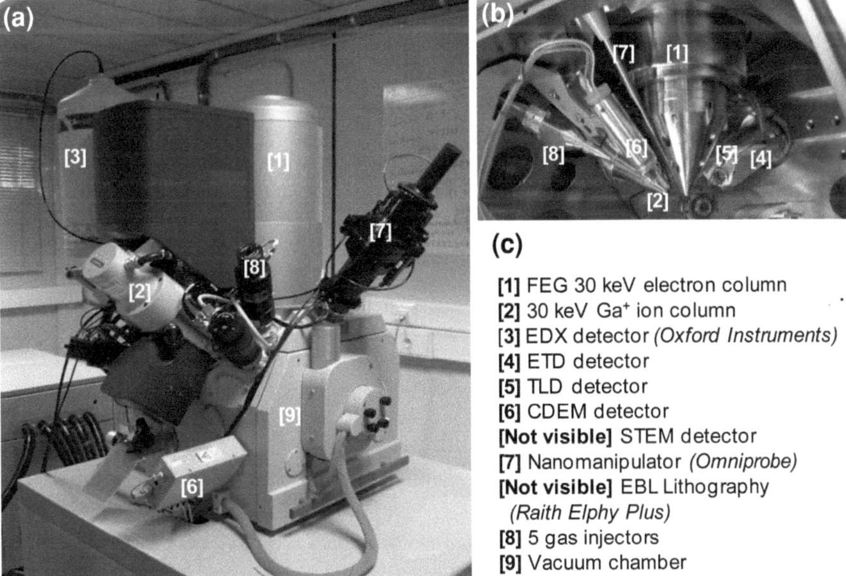

Fig. 2.5 Dual Beam equipment used. **a** General overview of the system. **b** View inside the chamber **c** Names corresponding to the numbers indicated in (**a**) and (**b**)

2.2 Electrical Measurements

2.2.1 Magnetotransport Measurements as a Function of Temperature

For the study of the electrical transport measurements of samples, as a function of temperature and magnetic field, we have used a combination of several items of equipment belonging to the ICMA (illustrated in Fig. 2.6):

- Combined Keithley® system, composed of a 6220 DC current source, and a 2182A nanovoltmeter. By injecting a constant current (DC), the voltage of the device under test is measured. A wide range of resistances can be measured, from 10 nΩ to 1 GΩ.
- Closed-Cycle-Refrigerator (CCR) by Cryocon®. By means of the thermodynamic cycle in He gas, the cryostat can lower the temperature of the sample down to 25 K. The control of temperature is performed by standard PID parameters.
- Electromagnet by Walker Scientific®, delivering maximum magnetic fields of 11 kOe.

A scheme of the system is shown in Fig. 2.7. The control of all equipment via PC was done by the design of Labview® programs. All this work was carried out jointly with Dr. Jan Michalik.

The electrical measurements are normally done in a 4-probe configuration, to avoid to measure contact resistances in the measurements. This geometry is schematized in Fig. 2.7.

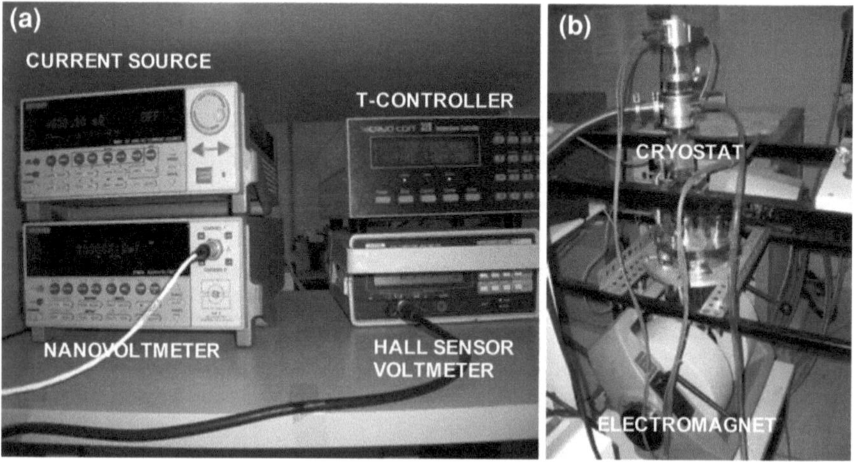

Fig. 2.6 Images of the system used for magnetotransport measurements as a function of temperature. **a** Equipments for electrical measurements, as well as for the control of temperature and magnetic field. **b** General view of the system

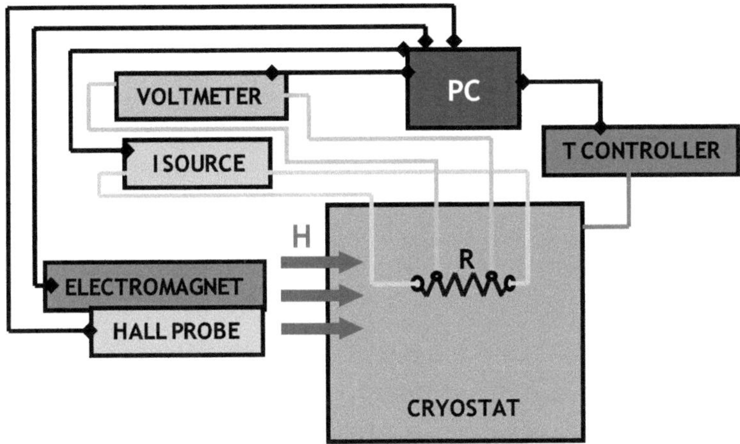

Fig. 2.7 Scheme of the system for magnetotransport measurements as a function of temperature

The Keithley system has been used in two modes of measurements:

- *Normal DC mode*: A constant current is applied, and voltage is measured.
- *Delta* mode: The voltage is measured with alternating positive and negative test current. This allows the cancelling of constant thermoelectric voltages by alternating the test current. It significantly reduces white noise, resulting in more accurate low resistance measurements when it is necessary to apply very low power. A few electrical measurements for this thesis were done in a commercial Physical Properties Measurement System (PPMS) from Quantum Design®, situated in the Instrumentation Service of the University of Zaragoza. This was used when high magnetic fields (maximum field of 9 T) or low temperatures (minimum temperature of 300 mK) were necessary, for resistances below ~ 1 MΩ.

2.2.2 *"In Situ" Electrical Measurements*

Some of the results obtained in the thesis have been achieved by a combination of the nanolithography techniques with electrical measurements. The usual approach followed consists of the study of the devices after its creation or modification. However, in this case we have also sometimes carried out simultaneous electrical measurements while the nanostructure was being patterned. To do this, two types of special stages were incorporated into the Dual Beam system (see Fig. 2.8):

1. Four Kleindiek® electrical microprobes, separately moved by a motor or piezoelectric, till contacted to pads. They are made of tungsten, and have a final diameter of around 1 μm.

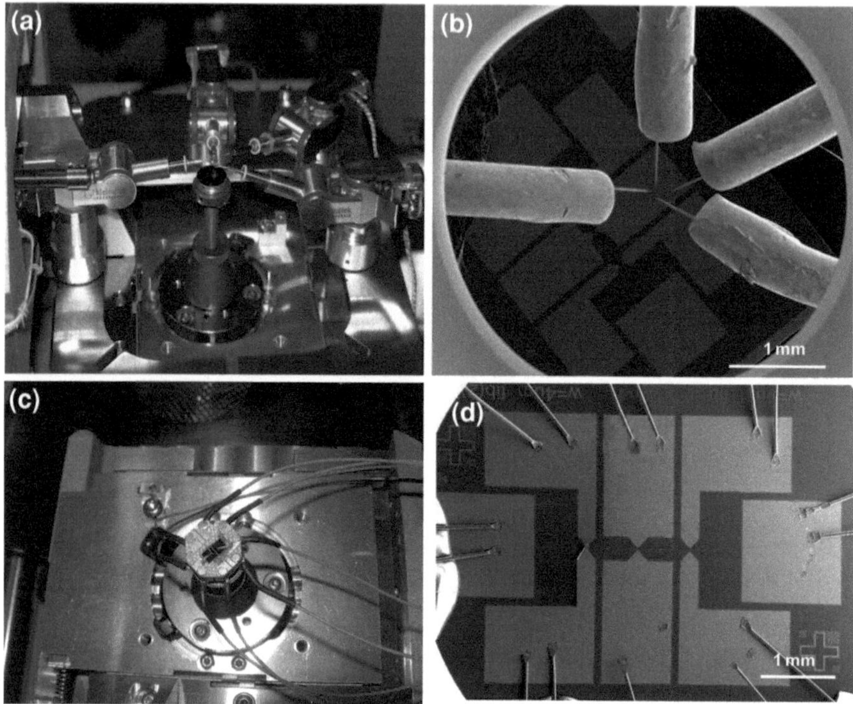

Fig. 2.8 Images of methods for electrical measurements inside the Dual Beam chamber. **a** Electrical microprobes stage. **b** SEM image of the microprobes inside the chamber, with a sample in a plane below. **c** Stage for electrical measurements. A sample is mounted in the image. **d** SEM image of a sample with microcontacts connected to the stage

2. Home-made stage, with eight connections, with the sample contacted by wire-bonding to a chip carrier, loaded into the stage.

The electrical wires are transferred outside the chamber by a port, and connected to the same Keithley system as explained before.

2.3 Spectroscopic Techniques

We have mainly used two different spectroscopy techniques to characterize he composition and chemical nature of the samples studied: energy dispersive X-ray spectroscopy (EDX) and X-ray photoelectron spectroscopy (XPS).

2.3.1 Energy Dispersive X-Ray Spectroscopy

We showed in Fig. 1.11a, that as a consequence of the interaction of electrons with matter in the keV range, X-ray photons are emitted. The PE produce the emission

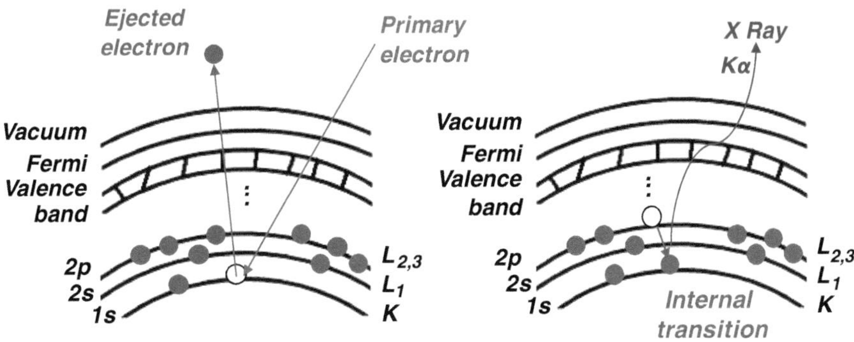

Fig. 2.9 Schematic diagram of the EDX process. Relaxation of the ionized atom results in photon radiation, characteristic for each element

of inner electrons, which create a hole in an atomic level. An electron from an outer, higher-energy shell can fill that hole, emitting an X-ray, with a characteristic energy equal to the difference between the higher and the lower energy shell. As the energy of the X-rays is characteristic of the difference in energy between the two shells, and of the atomic structure of the element from which they were emitted, this allows the elemental composition of the specimen to be measured. See a scheme of the process in Fig. 2.9.

An Oxford Instruments® EDX was used, incorporated into the Dual Beam system, for in situ compositional characterization (see Fig. 2.5). The interaction volume of the electron beam with the sample determines both the lateral resolution and the depth of analysis; this is a function of the primary beam energy, invariably of the order of one micrometer. The energy resolution is ~ 150 eV. This permits a compositional quantification to be made of the sample probed.

2.3.2 X-Ray Photoelectron Spectroscopy

XPS (traditionally called ESCA) is a spectroscopic technique based on the photoelectric effect, i.e., the ejection of an electron from a core level by an X-ray photon of energy $h\nu$. The energy of the emitted photoelectrons is then analyzed by an electron spectrometer. The kinetic energy (K) of the electron is the experimental quantity measured by the spectrometer, but this value will depend on the X-rays energy. The binding energy of the electron (BE) is the parameter which identifies an element specifically. The equation which describes the process is

$$BE = h\nu - K - W \tag{2.1}$$

where W is the spectrometer work function. Albert Einstein received the Nobel Prize in Physics in 1921 for his interpretation of the photoelectric effect. Figure 2.10 shows a scheme of the XPS process.

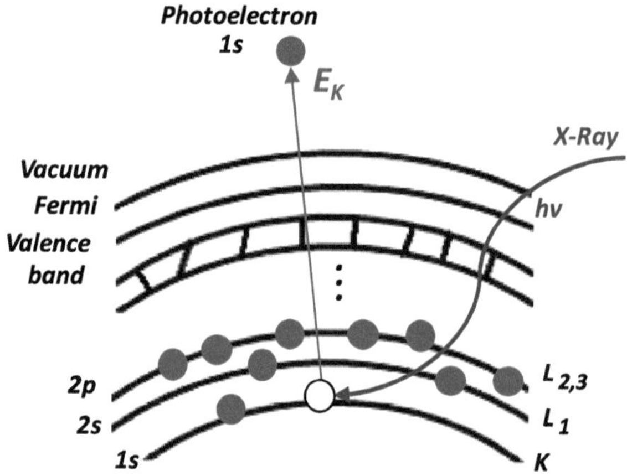

Fig. 2.10 Schematic diagram of the XPS process, showing the photoionization of an atom by the ejection of a 1 s electron

For these experiments we used a Kratos® Axis Ultra DLD equipment located at the INA. A monochromatic Al kα X-ray source was used, with exciting energy hv = 1486.6 eV. The sampling depth of the X-ray is at the micrometric level, while only those photoelectrons from the outermost, ~ 10 nm or less, escape without energy loss. These are the electrons used for quantitative quantification; XPS is thus a surface-sensitive technique. An in-depth study can be performed, since the equipment incorporates a 5 kV - Ar$^+$ gun, which progressively etches layers of material, which are consequently probed by XPS. A hemispherical detector analyses the electron energy. The experiments require ultra-high vacuum conditions (ionic pump).

Due to the difficulty in focusing X-rays, the effective minimum area probed in the equipment is approximately 30×30 μm^2. The high-resolution of the equipment (maximum = 0.2 eV) permits the valence state of the atoms composing the sample to be distinguished.

2.4 Spatially Resolved MOKE Magnetometry

The main magnetometry technique used in the thesis is a magneto-optic technique: the spatially resolved Longitudinal Magneto-Optical-Kerr-Effect (MOKE). These measurements were done at Imperial College, London, in the research group of Prof. Cowburn.

The Kerr effect consists of the rotation of the plane of polarization of a light beam when reflected from a magnetized sample. We show in Fig. 2.11. the geometry for the longitudinal Kerr-effect (L-MOKE). Radiation from a light source is first passed through a polarizer. The resulting plane-polarized light is then incident on a sample. The L-MOKE is sensitive to the in-plane magnetization component of the sample, and thus it is in the incidence plane of light. The magnetization changes the angle of polarization of light, as well as inducing an

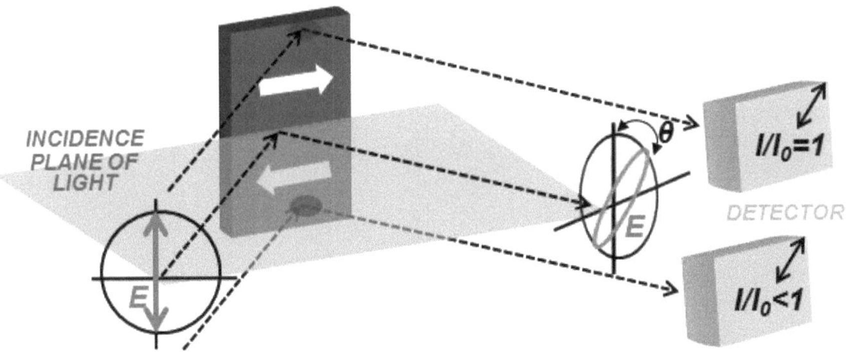

Fig. 2.11 Geometry for the Longitudinal-MOKE (see text). In the central ray of light, the polarization of the electric field changes from linear to elliptical polarization, with a change in the angle of polarization, θ. In the top ray, the detector is aligned to make the signal maximum. Thus, due to the contrary angle rotation in the bottom ray, resulting from an opposite magnetization direction in the sample, the signal collected by the detector is smaller

ellipticity in the initially linearly polarized light. The thickness probed is typically of 20 nm. In the example of Fig. 2.11, the sample contains two domains magnetized in opposite directions. The light incident on one domain is rotated in the opposite direction from that incident on the other domain. Therefore, if the analyzer is oriented such that the light from the first domain is maximum, then the plane of polarization of the light reflected from the other domain is not aligned with the analyzer, and the signal is reduced.

A simple explanation for this effect is as follows [7]. Linearly polarized light can be decomposed in two oppositely circular polarizations. The angular number of both is equal to 1, but in the right circular polarization (+), $m_L = +1$, whereas the left circular light (−) has $m_L = -1$. Using the simple example of a magnetic material with atomic spin $S = 1/2$, the exchange interaction splits its energy levels into two sub-levels, with total spin $m_S = +1/2$ and $m_S = -1/2$.

Both energy and angular momentum must be conserved when a photon excites an electron from one sub-level in A to one in B. Selection rules dictate that $\Delta m_L = \pm 1$, implying that only transitions drawn in Fig. 2.12 are possible. Thus, oppositely polarized photons correspond to different electronic transitions in the atom. If the electronic population in B sub-levels differ one from the other, the absorption of one polarization is greater than the other (a phenomenon called circular dichroism). When the resulting circular polarizations are recombined again, the plane of polarization is rotated with respect the incoming beam. The resulting phase difference between the initial and final planes of polarization is called circular birefringence.

The magneto-optical effects do not directly provide absolute values for the magnetization. The MOKE rotation depends on the angle of measurement and wavelength of light, as well as on the magneto-optic constant, which is material dependent at a determined temperature. Thus, MOKE measurements provide

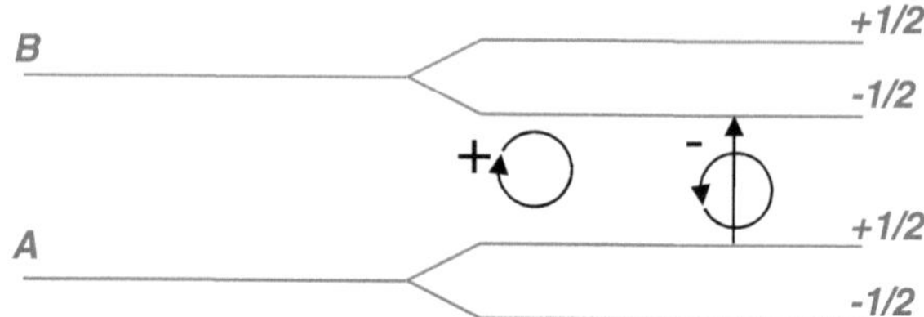

Fig. 2.12 Absorption of light in a ferromagnet with S = 1/2

hysteresis loops in intensity (arbitrary units) as a function of the magnetic field. The Kerr effect is normally used for measurements of hysteresis loops in magnetic thin film layers, or for the imaging of magnetic domains. However, this experimental setup [8, 9], now commercialized by Durham Magneto Optics®, is suitable for the measurement of magnetic nanometric structures. A diode-pumped solid state laser ($\lambda = 532$ nm) with a diameter of about 5 μm (FWHM) is used as a probe. A CCD camera is used for pre-alignment of the nanometric structure, located on a substrate. The final alignment is done by performing a reflectivity map of the zone of interest. A motor permits the stage to be moved with a resolution of 50 nm. A quadruple is used for applying AC magnetic fields (1–27 Hz), with H_{max} ~ 400 Oe. The system is a state-of-the-art-MOKE instrument having, for instance, a sensitivity for Permalloy $S = 6 \times 10^{-12}$ emu at room temperature [9] (in a SQUID magnetometer $S \sim 10^{-8}-10^{-9}$ emu). This high sensitivity permits the measurement of hysteresis loops in sub-micrometric and sub-λ nanomagnets. As the diameter of the laser is bigger than at least one of the dimensions of the nanostructure probed, a dilution factor is present in the MOKE signal with respect to the bulk material.

2.5 Atomic Force Microscopy

In the previous sections we have cited two main microscopy techniques used for this work: the optical microscope, allowing the observation of structures with sizes slightly below 1 μm, and the scanning electron microscope, which can resolve objects around 1 nm in size.

In this section, we will comment on some aspects of another microscope used, with resolution of fractions of a nanometer: the atomic force microscope (AFM). AFM is a type of scanning probe microscope (as is its parent, the scanning tunneling microscope), based on the interaction of a sharp tip with the surface sample when it is brought into proximity to the surface. The tip is the end of a microscopic cantilever, used to scan the specimen surface. The cantilever is typically of silicon or silicon nitride with a tip radius of curvature in the order of nanometers. The forces between the tip and the sample lead to a deflection of the cantilever, which can be modeled with Hooke's law. The typical AFM forces are mechanical contact force, Van der Waals force, capillary forces, electrostatic forces, magnetic forces

(MFM), etc. Typically, the deflection is measured using a laser spot reflected from the top surface of the cantilever into an array of photodiodes.

The AFM images were taken at Imperial College, London, with a Veeco Multimode® AFM. The Ph.D student Liam O'Brien made a part of the measurements. Contact mode was used for imaging, where the force between the tip and the surface is kept constant during scanning by maintaining a constant deflection. The tip deflection is used as a feedback signal, which is used to form an image of the structure probed.

Some clear advantages of AFM compared to SEM can be mentioned. There is better spatial resolution, the measurements do not require vacuum conditions, a true three-dimensional surface profile is obtained, and there are no charging-effect problems with insulator samples. However, there are also important disadvantages of AFM compared with SEM. The SEM can image an area on the order of mm × mm, with a depth of field in the order of mm. The AFM can only image a maximum height in the order of microns, and a maximum scanning area of a few μm^2. Besides, image artifacts are common, especially if an incorrect choice of tip for the required resolution is made. The significant lower speed in the image is also an important drawback, as well as the possible hysteresis in the piezoelectric. We should mention for the sake of completeness that AFM is also used as a tool for nano-patterning, by local oxidation, dip-pen nanolithography, etc.

2.6 High Static Magnetic Fields

The high static magnetic field experiments were performed at the High Field Magnet Laboratory (HFML) of the University of Nijmegen, Netherlands. The high fields are generated by a Bitter type magnet, consisting on stacked copper (no magnetic material) disks, with very high current densities flowing in them. Small vertically aligned holes are pierced to allow cooling by a high pressure water flow (see Fig. 2.13a). Static fields as high as 33 T can be generated at maximum power of 20 MW (40 kA at 500 V). This is the technique nowadays that produces the highest static magnetic fields (we can compare it with the maximum field produced by: a solenoid without cooling ~ 0.1 T, an electromagnet ~ 2–3 T, or a super-conducting Nb_3Sn coil at 4.2 K ~ 20 T [10]). Fig. 2.13b shows an experimental station of the Laboratory, where the coils are inside a cryostat for measurements as a function of temperature. Dr. Uli Zeitler and Erik Kampert were our local contacts for these experiments.

2.7 Other Techniques

Some results in this thesis will be presented with other techniques different from those previously explained, which should also be mentioned. The person in charge

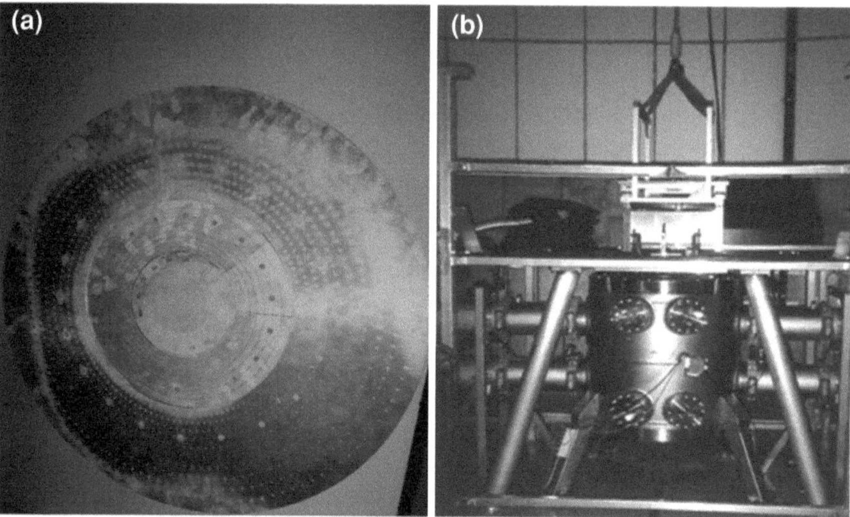

Fig. 2.13 **a** Bitter coil. **b** Experimental station at the HFML

of the measurements is cited in the text when a particular result is presented in the following chapters.

Pulsed Laser Deposition, PLD (INA): Fe_3O_4 thin films were epitaxially grown on MgO (001) substrates. A KrF pulsed laser produces an ablation in a Fe_3O_4 target, depositing stochiometrically the iron oxide on the substrate.

High-Resolution X-ray Diffraction, HR-XRD (INA): structural characterization of thin films. Determination of the degree of crystallinity of a film grown on a substrate.

Superconducting quantum interference device, SQUID magnetometer (Scientific Services of University of Zaragoza): a constant current passes through two in-parallel Josephson junctions. The applied voltage oscillates with the changes in the phase of the two junctions, which depend upon the change in magnetic flux. This results in a highly sensitive magnetometer.

High-Resolution Transmission Electron Microscope, HRTEM (Scientific Services of the University of Barcelona, CEMES-Toulouse): 200–300 kV electrons are accelerated, being transmitted through a thin sample ($\sim$ 100 nm). The image formed, a consequence of the diffraction of electrons with the material, can be transformed to the real space, resulting in an image of the crystallographic structure of a sample at an atomic scale.

Scanning-Tunneling-Spectroscopy, STS (University Autonóma of Madrid, group of Prof. Sebastián Vieira): the local density of electronic states of a surface is probed at an atomic scale, by measuring the tunneling differential conductance between a STM tip and a sample at low temperatures.

References

1. For details of the growth process of the Fe_3O_4 films by PLD, contact: juliao@unizar.es
2. L.R. Harriott, Limits of lithography. IEEE Proceedings (2001)
3. T. Ito, S. Okazaky, Pushing the limits of lithography. Nature **406**, 1027 (2000)
4. H.J. Levinson, *Principles of Lithography* (SPIE Press, Bellingham, 2005)
5. H.J. Levinson, M.A. McCord, F. Cerrina, R.D. Allen, J.G. Skinner, A.R. Neureuther, M.C. Peckerar, F.K. Perkins, M.J. Rooks, in *Handbook of Microlithography, Micromachining, and Microfabrication*, ed. by P. Rai-Choudhury (SPIE Press Monograph vol. PM39, 1997)
6. L. Giannuzzi, F. Stevie, *Introduction to Focused Ion Beams: Instrumentation, Techniques, Theory and Practices* (Springer Science + Business Media, Boston, 2005)
7. N. Spaldin, *Magnetic Materials, Fundamentals and Device Applications*. Chapter 12 (Cambridge University Press, Cambridge 2003)
8. R.P. Cowburn, D.K. Koltsov, A.O. Adeyeye, M.E. Welland, Probing submicron nanomagnets by magneto-optics. Appl. Phys. Lett. **73**, 3947 (1998)
9. D.A. Alwood, Gang. Xiong, M.D. Cooke, R.P. Cowburn, Magneto-optical Kerr effect analysis of magnetic nanostructures. J. Phys. D Appl. Phys. **36**, 2175 (2003)
10. È. Du Trémolet de Lachisserie, D. Guignoux, M- Schneider. Magnetism: vol II: Materials and applications. Chapter 26 (Springer, Heidelberg 2003)

Chapter 3
Magnetotransport Properties of Epitaxial Fe₃O₄ Thin Films

In this chapter we will show the magnetotransport properties of epitaxial Fe_3O_4 thin films [grown on MgO (001)]. A broad characterization has been performed, studying systematically, for several film thicknesses and as a function of temperature, the resistivity, magnetoresistance in different geometries, Hall effect and planar Hall effect. Most of the measurements are substantially influenced by the presence of the antiphase boundaries, especially for ultra-thin films. In the case of the anomalous Hall effect, experimental evidence of the universal character of this phenomenon was demonstrated.

The growth of the films was conducted by J. Orna, this being the subject of her Ph. D. Thesis.

3.1 Introduction

3.1.1 General Properties of Fe₃O₄

Magnetite, Fe_3O_4 is the first magnetic material referred in the history of human-kind. Its magnetic properties were reported in the *Chinese book of the Devil Master*, in the fourth century BC. This iron oxide is the magnetic mineral contained in lodestone, from which the first compasses for navigation were made. In spite of this long history, many aspects of the physics of magnetite remain unresolved.

Magnetite is magnetic at room temperature, being ferrimagnetic, with a Curie temperature of 858 K [1]. It belongs to the cubic ferrite family: $Fe^{2+}O^{2-}$ · $(2Fe^{3+}3O^{2-})$, with an inverse spinel crystallographic structure (lattice constant $= 8.397$ Å). This structure (Fd3m space group) consist of a face-centered-cubic (FCC) oxygen lattice, with Fe^{3+} ions filling 1/8 of the available tetrahedral

A. Fernandez-Pacheco, *Studies of Nanoconstrictions, Nanowires and Fe₃O₄ Thin Films*, Springer Theses, DOI: 10.1007/978-3-642-15801-8_3,
© Springer-Verlag Berlin Heidelberg 2011

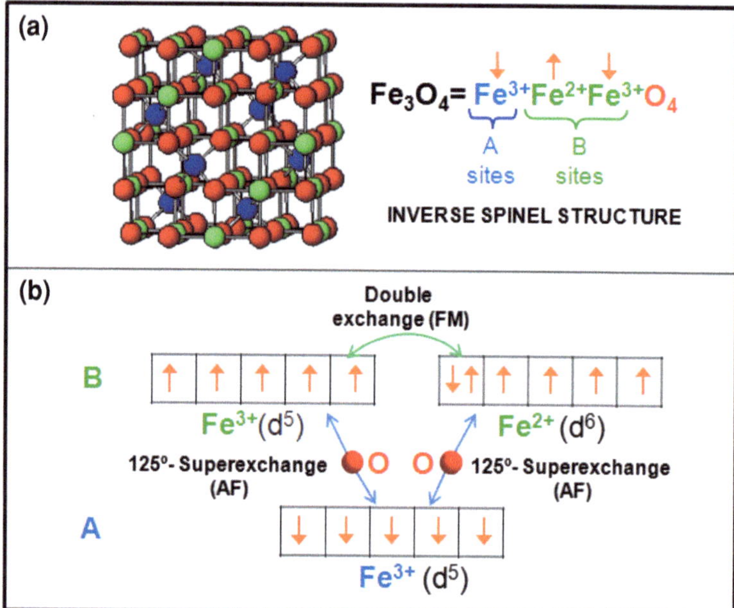

Fig. 3.1 **a** The inverse spinel structure of Fe$_3$O$_4$, consisting of a FCC oxygen lattice, with octahedral (*B*) and tetrahedral (*A*) sites, filled by Fe^{2+} and Fe^{3+} cations (see text for details). **b** Scheme of the exchange interactions present in magnetite. The strong antiferromagnetic (AF) superexchange interaction between *A* and *B* ions, mediated by oxygen ions, forces the parallel alignment of ions in *A* sites. The ferromagnetic (FM) double-exchange interaction between Fe^{2+} and Fe^{3+} permits the hopping of electrons between *B* sites

sites (A sites), and equal amounts of Fe^{2+} and Fe^{3+} ions filling half of the available octahedral site (B sites), see Fig. 3.1a.

The spin moments of all the Fe^{3+} ions on B sites are ferromagnetically aligned and in the opposite direction to the Fe^{3+} occupying the A sites. Therefore the magnetic moments of all Fe^{3+} cancel out, resulting in a zero contribution to the overall magnetization of the solid. However, all the divalent ions are ferromagnetically aligned in parallel, providing a net magnetization of approximately 4 μ_B/f.u. Several magnetic exchange interactions are present in Fe$_3$O$_4$, which are detailed in Fig. 3.1b.

The conduction in magnetite has been extensively studied and modeled. Put in simple terms, as a consequence of the double-exchange (FM) interaction existing between Fe^{2+} and Fe^{3+} in B sites, the additional spin-down electron can hop between neighboring B-sites, resulting in a high conductivity, and an intermediate valence of Fe$^{2.5+}$. At around 120 K (T$_v$), a first-order phase transition occurs, the so called Verwey transition [2]. At this transition, the structure distorts from cubic symmetry [3, 4], and a charge ordering occurs at the B sites, reducing the conductivity by two orders of magnitude [3, 4]. Conduction is, however, much more

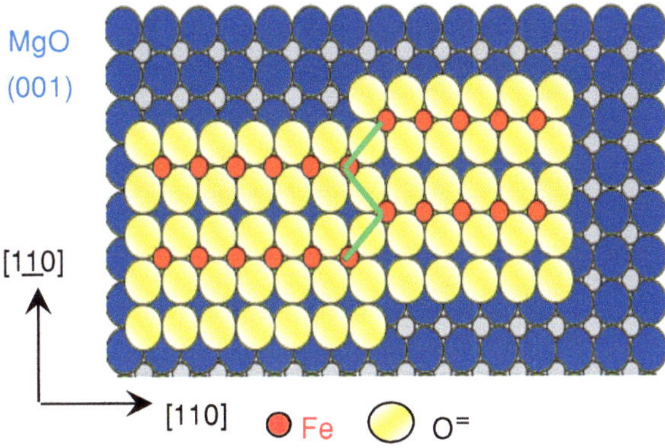

Fig. 3.2 Schematic representation of an APB in Fe_3O_4 formed in a MgO (001) substrate. When right and left iron oxide islands coalesce while growing, a loss of periodicity can occur, forming a boundary

complex. We could cite the models developed by Ihle and Lorentz [5] and Mott [6] as currently being the best established.

In accordance with the qualitative description for conduction explained, band calculations [7, 8] predict magnetite to be a half metal, with $P = -1$ above T_v, making it a promising material in MTJs devices (see Sect. 1.4).

3.1.2 Properties of Epitaxial Fe_3O_4 Thin Films

Many groups have worked during recent years on growing epitaxial Fe_3O_4 thin films by MBE and PLD [9–17]. The most commonly chosen substrate is single crystal MgO (001), because its cubic lattice parameter, 4.21 Å has a small lattice mismatch with bulk Fe_3O_4, permitting the epitaxial growth of films, making Fe_3O_4 a promising material to form magnetic-tunnel-junctions electrodes (see Sect. 1.4).

The most important aspect to remark in Fe_3O_4 thin films with respect to the bulk material is the presence of structural defects, caused by the growth mechanism, called antiphase boundaries (APBs). APBs form when islands of Fe_3O_4 coalesce and neighboring islands are shifted with respect to each other. In Fig. 3.2 a scheme of an APB is shown. There exist different boundaries, depending on the vector shift present. For details of APBs see Refs. [11, 18].

These structural defects play a major role in the properties of magnetite thin films, mainly due to the superexchange AF interactions between spins surrounding the boundary. The important increase of resistivity [11], the lack of saturation in the magnetization [17], or a higher magnetoresistance (MR) effect than in single crystals [10, 16, 17], as a consequence of the spin-polarized transport, are caused by APBs. It is important to point out that spin-resolved photoemission studies in thin films show that APBs do not seem to reduce significantly the spin polarization in films [19].

3.2 Experimental Details

3.2.1 Growth of the Films

The Fe$_3$O$_4$ target was prepared from Fe powder by solid-sate reaction [20]. Fe$_3$O$_4$ thin films were grown on MgO (001) substrates by PLD using a KrF excimer laser with 248 nm wavelength, 10 Hz repetition rate and 3×10^9 W/cm^2 irradiance. The base pressure in the deposition chamber was lower than 10^{-8} Torr, and the substrate was kept in vacuum at $\approx 400°$C during laser ablation of the target.[1]

3.2.2 Types of Electrical Measurement: Van Der Pauw and Optical Lithography

The magnetotransport measurements were performed in the combined CCR-electromagnet-Keithley system explained in Sect. 2.2. Basically, all the work that will be explained below comprises two types of measurements:

- As a function of the thin film thickness, at room temperature: 5×5 mm^2 samples were deposited, being measured by the Van Der Pauw technique [21] (current // [001] or // [110], depending on whether longitudinal or transversal resistivity components are measured) after wire-bonding to the corners of the sample (see Fig. 3.3a$_1$, a$_2$).
- As a function of temperature, for selected thicknesses $= 40$ and 20 nm: an optical lithography process was carried out in 0.5×0.5 inch^2 samples (current // (001), see Fig. 3.3b and Sect. 2.1.1), since the high resistance values for Fe$_3$O$_4$ at low temperatures (especially below T$_v$) would make the measurements of transversal voltages (Hall and planar Hall effect) unfeasible with the Van Der Pauw method.

3.3 Structural and Magnetic Characterization

Together with the thorough transport characterization explained in the following sections, the films were characterized structurally and magnetically[2] to demonstrate their epitaxy.

---

[1] For details of the growth process of the Fe$_3$O$_4$ films by PLD, contact: juliao@unizar.es.
[2] See Footnote 1.

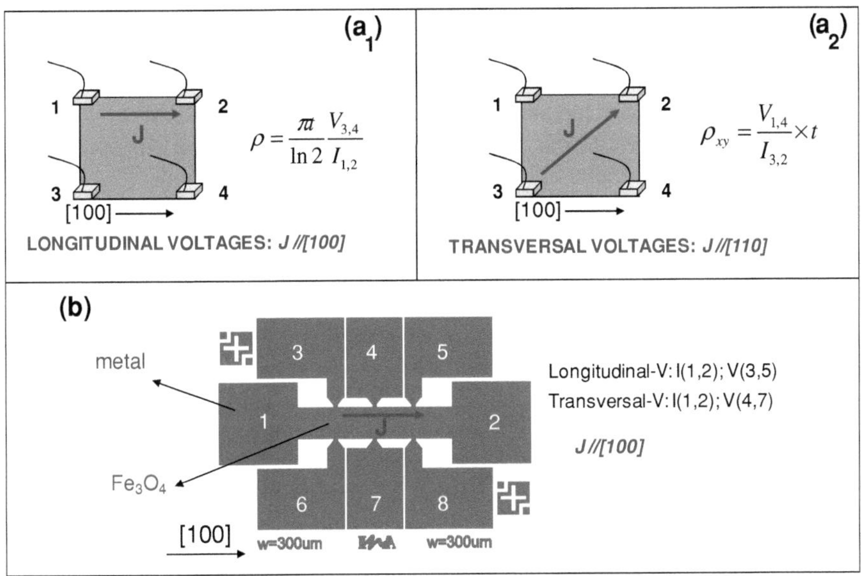

Fig. 3.3 Geometry for the two types of measurements done in Fe_3O_4 thin films. **a** Van Der Pauw measurements at room temperature, as a function of film thickness. Configuration for $\mathbf{a_1}$ longitudinal and $\mathbf{a_2}$ transversal voltages. The explicit expressions for the components of the resistivity tensor are shown. **b** Sample for temperature-dependent measurements. Several equivalent pads were designed, for more flexibility in the measurements. Metal was deposited on these pads, on top of magnetite, to minimize the contact resistances

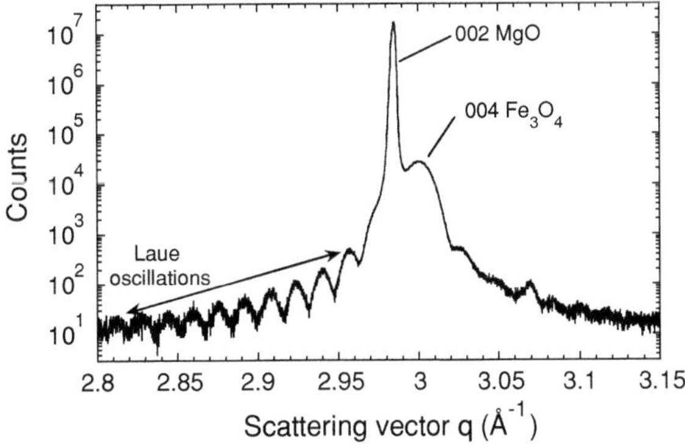

Fig. 3.4 Symmetrical $\theta/2\theta$ scan as a function of the scattering vector q, around the 002 Bragg peak from the MgO substrate, showing the 004 reflection from the Fe_3O_4 film along with its Laue oscillations. The small peak at $q = 3.07$ Å^{-1} is a reflection from the silver rest used to fix the substrate during deposition. Experiments performed by J. Orna and J.A. Pardo

In Fig. 3.4 we show X-ray diffraction data taken on a 40 nm thick film. The symmetrical $\theta/2\theta$ scan around the 002-Bragg peak from the MgO substrate shows the 004-reflection from the Fe_3O_4 film, which gives the expected orientation Fe_3O_4

(001) // MgO (001). The Fe$_3$O$_4$ out-of-plane lattice parameter extracted from the peak position is $a_{out} = 8.38$ Å. This value, lower than the bulk Fe$_3$O$_4$ lattice parameter, indicates that the film is tensile strained in the plane, due to the slight differences in the lattices parameters between film and substrate, and it is therefore compressively strained in the out-of-plane direction. Moreover, the periodicity Δq of the observed Laue oscillations gives a coherence length $\xi = 2\pi/\Delta q = 39 \pm 1$ nm, which indicates the existence of crystalline coherence along the full thickness of the film. The observation of Laue oscillations, up to 10th order, is normally taken as an indication of a very high crystalline coherence.

Magnetic measurements in a SQUID magnetometer revealed saturation magnetization at room temperature of $M_s = 440$ emu/cm^3. As commented previously, the decrease of M_s with respect to bulk ($M_s = 480$ emu/cm^3) is associated to the AF coupling in APBs [17].

Direct visualization of the APBs network in our samples was achieved through plane-view TEM measurements. The TEM work was done by César Magén (cmagen@unizar.es) and Ettiene Snoeck in the CEMES-CNRS laboratory in Toulouse (France). Figure 3.5 shows the dark-field image of a plane-view TEM specimen (40 nm thick film) collected by selecting a (220)-type reflection. The APBs are located where there is darker contrast. The mean size of domains between boundaries (APDs: Antiphase domains) for this sample is 24.4 nm with a standard deviation of 13.9 nm. This average domain size is comparable to other reported results for this film thickness [10, 11].

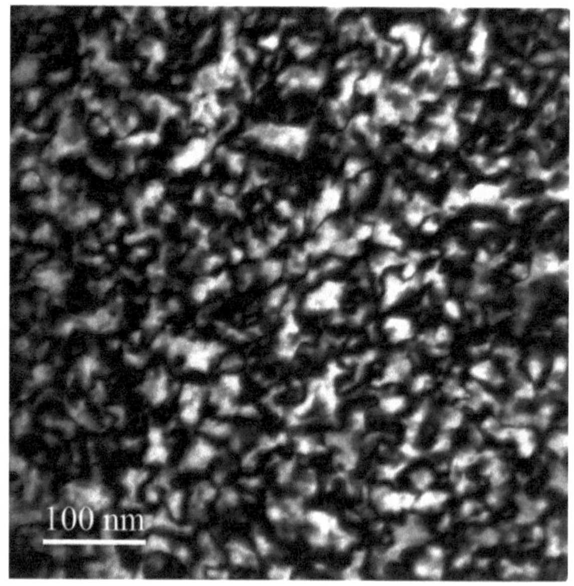

Fig. 3.5 300 kV TEM planar-view image of a 40 nm-thick film, where the APB network can be observed. The (220)-type reflection has been selected close to a two-beam condition, in which only the direct beam and one (220)-type reflection are strongly excited, near the [001] zone axis. Image taken by C. Magén at CEMES, Toulouse

3.4 Resistivity

3.4.1 Resistivity as a Function of the Film Thickness (Room Temperature)

The longitudinal resistivity ($\rho_{xx} \equiv \rho$) values at room temperature for films of thickness (t) ranging from 150 down to 4 nm are shown in Fig. 3.6, compared with other works in the literature [9–12, 16]. The good accordance in the data confirms the high quality of the films. The resistivity enhancement as the films get thinner, apart from the normal increase due to surface scattering, is a direct consequence of the increase in the density of APBs ($\rho \propto \text{thickness}^{-1/2}$) [10, 11]. The thinnest film studied, t = 4 nm, has a resistivity $\sim$40 times higher than the bulk, of 4 mΩ cm.

3.4.2 Resistivity as a Function of Temperature

Selected samples were measured as a function of temperature. Semiconducting behavior of the resistivity is observed in the film as a function of temperature, as occurs in the bulk samples (see Fig. 3.7). The Verwey transition is observed as a significant increase of ρ. As the films get thinner, the transition becomes smoother and smoother, becoming imperceptible for t = 5 nm. The Verwey temperature for films is lower than in single crystals ($T_V \sim 120$ K). The decrease of T_V with

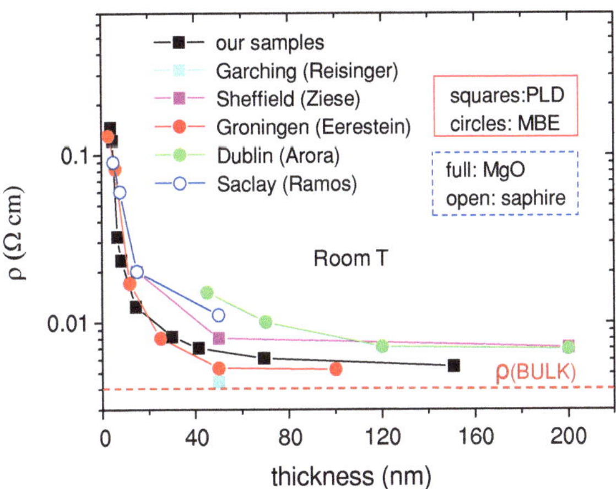

Fig. 3.6 Rom temperature resistivity measurements over a wide range of film thickness. Our data are compared with several groups (Reisinger et al. [9], Ziese et al. [16], Eerestein et al. [11], Arora et al. [12] and Ramos et al. [10]), in good agreement. The data symbolized as *squares* or *circles* refer to samples grown by PLD or MBE, respectively. All these films were grown on MgO, except for the work in Saclay, where they used *sapphire*

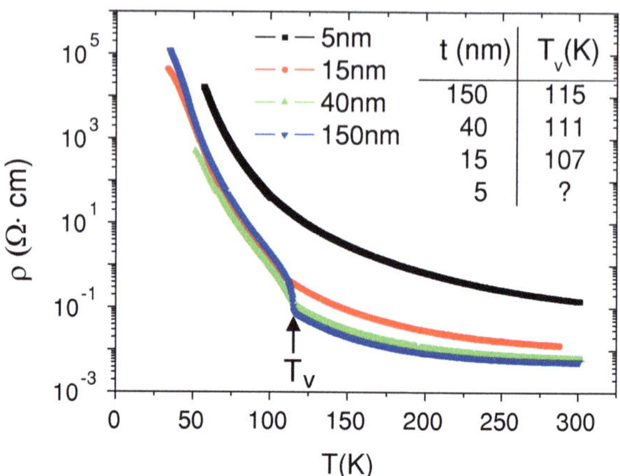

Fig. 3.7 Resistivity as a function of temperatures for four thicknesses. The Verwey transition is indicated with an arrow. As the films get thinner, the transition becomes less abrupt, and T_v is shifted towards lower temperatures

respect to the bulk value has been observed in epitaxial thin films for thicknesses below 100 nm. The two effects observed as films get thinner, the decrease of T_v and disappearance of the transition have been associated to the stress caused by the substrate [22] and the inhibition of long range order for very thin films, as a consequence of the high density of APBs [11, 23].

3.5 Magnetoresistance and Anisotropic Magnetoresistance

As was explained in the introductory section, epitaxial Fe_3O_4 thin films present higher (although moderate) magnetoresistance (MR) values than single crystals, due to the spin-polarized transport across sharp antiferromagnetic boundaries [16, 18]. We have used the convention, for the definition of MR:

$$MR(\%) = 100\frac{[(R(H) - R_0)]}{R_0} \tag{3.1}$$

3.5.1 Geometries for MR Measurements

Using the CCR equipment detailed in Sect. 2.1.1 we have measured the MR up to 11 kOe in three geometries, depending on the relative direction of the magnetic field H with respect to the current and substrate plane (see Fig. 3.8): H perpendicular to the thin film plane (perpendicular geometry: PG), H in plane and parallel to the current I (longitudinal geometry: LG), H in plane and perpendicular to I (transversal geometry: TG). Measurements in the PG were also done for a 40 nm thick film up to 90 kOe in a PPMS by Quantum Design, as well as at the HFML in Nijmegen applying fields up to 300 kOe (Sect. 2.6).

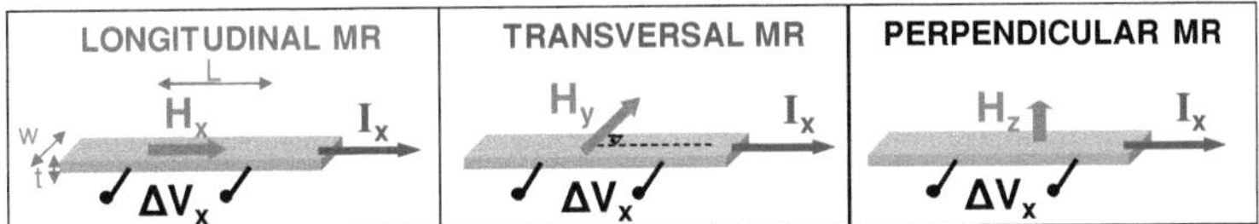

Fig. 3.8 MR geometries $\left(\rho_{xx} = \frac{\Delta V_x}{I_x}\frac{wt}{L}\right)$

3.5.2 MR as a Function of Film Thickness (Room Temperature)

In Fig. 3.9 a typical example of the measurements performed up to 11 kOe is shown for a 40 nm thick film, for the three geometries. The magnetoresistive effect, in all cases, is small (a few percent) and negative. The shape of the MR curves is linear for the LG and TG in the range of fields measured. For the PG, the shape is parabolic and linear, for fields lower and higher than the anisotropy field, respectively. Reference [18] models this behavior.

In Fig. 3.10 the MR value in the LG, at maximum H = 11 kOe, is shown for all the thicknesses studied, together with the coercive field (H_c) derived from the measurements (field where MR presents peaks, see Fig. 3.11). As the films become thinner, the MR first increases, reaching a maximum at around 40 nm, and decreases abruptly for t < 30 nm. The increase of MR as t decreases can be understood by the presence of a higher density of APBs in the films, since the MR effect is mainly due to the transport though them. However, the decrease in the size of the APDs for thinner films results in a tendency of the domains to behave superparamagnetically [24], decreasing the MR effect. This superparamagnetic behavior is manifested by the evolution of H_c with thickness, and has been widely studied in planar Hall measurements (see Sect. 3.6)

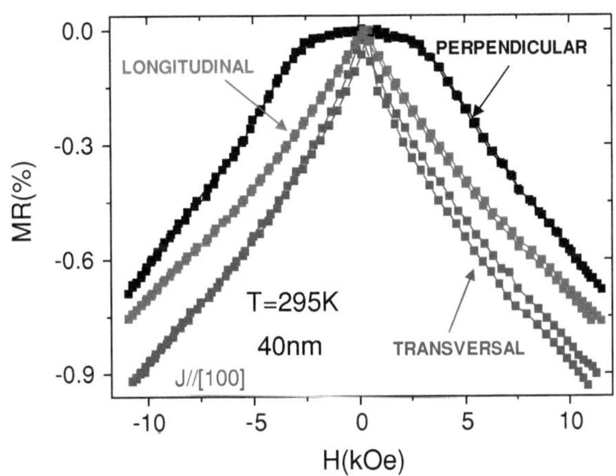

Fig. 3.9 MR measurements for a 40 nm thick film, at room temperature, in three geometries. The MR effect is mainly caused by the spin-polarized transport though APBs

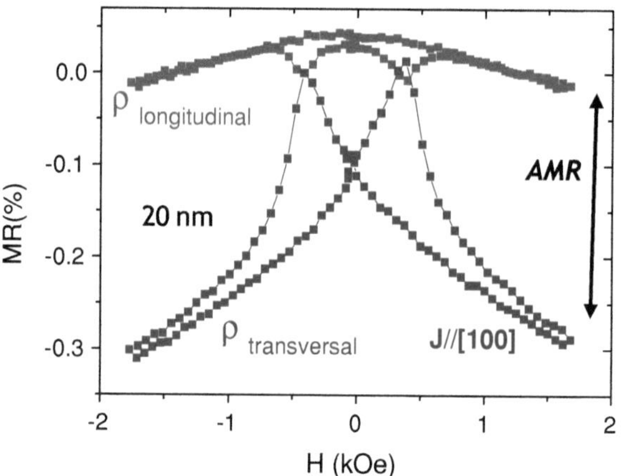

Fig. 3.10 MR value at 11 kOe in the LG for all the films studied, at room temperature (*squares*). The evolution of MR with thickness is well correlated with the behavior for H$_c$ (*circles*). Both behaviors are understood by the increase of the density of APBs as the films get thinner. *Lines* are guides to the eye

Fig. 3.11 MR measurements at room temperature for a 20 nm thick film. The anisotropic MR effect is evidenced, with AMR > 0 in this case. Peaks in the MR are associated to the coercive field

In Fig. 3.11 the MR measurements for a 20 nm thick film, at room temperature, are shown for a low range of fields (-1.5 kOe $<$ H $<$ 1.5 kOe). The difference in resistance between the LG and the TG is the AMR. The AMR, previously introduced (Sect. 1.6.2.3) is defined in Eq. 1.8, with $\rho_\parallel$ and $\rho_\perp$ the resistivities when current and magnetization are parallel (LG) and perpendicular (TG), respectively, and ρ_0 is the resistivity at the coercive field.

The origin of the AMR lies in the spin–orbit coupling, and was first explained by Kondo in the early-1960s [25]. Briefly, the s electrons, which are responsible for the conduction, are scattered by the unquenched part of the orbital angular moment of the 3d electrons. As the spin magnetization direction rotates in response to the applied magnetic field, the 3d electron cloud deforms, due to the spin–orbit coupling, and changes the amount of scattering of the conduction electrons. The process is schematically shown in Fig. 3.12. In magnetite, the AMR sign depends on the direction of the current with respect to the crystallographic direction [16]. AMR has been studied in detail in the PHE measurements (Sect. 3.6).

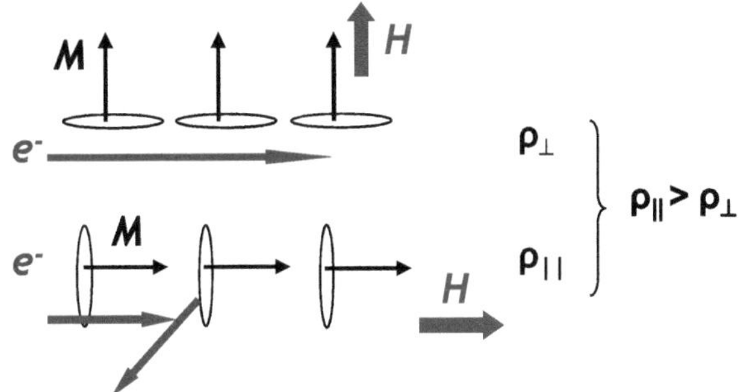

Fig. 3.12 Schematic representation of the origin of AMR

3.5.3 MR as a Function of Temperature

A lot of work has been devoted to the study and modeling of the MR in epitaxial Fe_3O_4 thin films under moderate magnetic fields [10, 15, 16, 18], in a range where the films are not magnetically saturated due to the AF coupling between spins at the APBs (Sect. 3.1.2). It has been well established that the spin-polarized transport of electrons through the APBs is the leading physical mechanism behind this negative MR, substantially higher than in single crystals [16]. The novel contribution in this thesis consists on the measurements of the MR applying very high static magnetic fields, up to 30 T.

In Fig. 3.13 we show the MR in the PG for a 40 nm-thick sample, below room temperature (down to 80 K, including $T < T_v$). For $T > T_v$ we fitted our MR curves to the model developed in Refs. [10, 18] for polarized transport through APBs:

$$MR(\%) = -\left[C\frac{M_\infty}{W_{AF}}\mu_0 H - \left(\frac{M_\infty}{W_{AF}}\mu_0 H\right)^{3/2} \right] \tag{3.2}$$

(where μ_0 is the vacuum permeability, and C and M_∞/W_{AF} are fitting constants). The fit is good for fields in the range $H_a < H < H_{max}$, H_a being the anisotropy field, since for $H < H_a$, MR is quadratic [18]. H_{max} is slightly temperature-dependent, around 8–10 T. For example, fitting values at 242 K are $C = 35.0 \pm 0.1$ and $M_\infty/W_{AF} = (0.03 \pm 10^{-4})$ T^{-1} in good agreement with results reported by Ramos et al. [10], who studied MR up to 7 T. For $H_{max} < H < 30$ T, the MR is not saturated, and becomes more linear with field than the prediction of this model (MR would tend more to saturation), thus Eq. 3.2 cannot be applied. For $T < T_v$, the MR changes its shape, and this model is also not appropriate. We have also fitted the MR to the model used in Ref. [26] for Zn-doped-Fe_3O_4 up to 15 T, previously introduced for perovskites manganites [27], where the MR is explained by the spin-polarized hopping of localized carriers in presence of magnetic disorder. In this case the MR follows the Brillouin function B_J:

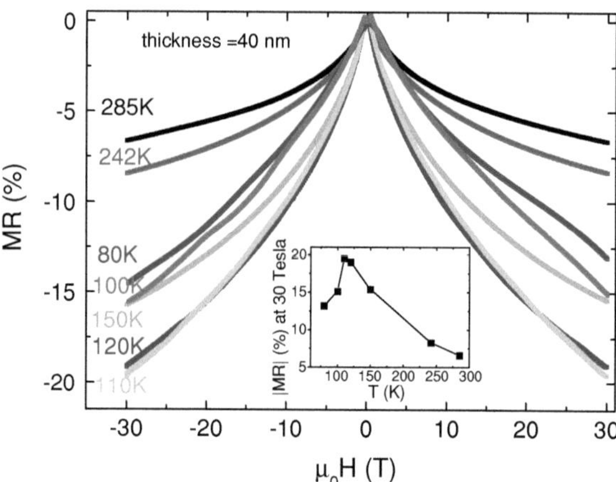

Fig. 3.13 MR up to 30 T measured in a lithographied 40 nm-thick sample, for temperatures above and below the Verwey transition, with the magnetic field applied perpendicular to the substrate plane. A linear, intrinsic contribution is found in addition to the expected behavior by the antiphase boundaries. This tendency, manifested at high fields, is of special relevance below the transition. In the inset, the MR (in absolute value) at 30 T is plotted versus temperature. A maximum value of 20% is found at T$_v$

$$MR(\%) \approx -\beta \frac{\mathbf{B}_J\left(\frac{\mu_{eff}\mu_0 H}{k_B T}\right)}{k_B T} \tag{3.3}$$

where μ_{eff} is the effective magnetic moment of the local spins, and J = 2. The accordance with the model is good at T > T$_v$, for fields lower than $\sim$ 15 T, and we obtain $\mu_{\text{eff}} \approx 90 \ \mu_B$, in good agreement with Ref. [27]. (MR would be understood by small misalignments of spins between $\sim$3 unit-cell-sized-clusters). The fit is suitable in the complete field range for T < T$_v$, where the conduction becomes more localized, by a variable-range-hopping mechanism [3, 6], yielding $\mu_{\text{eff}} \approx 10 \ \mu_B$. This value is substantially smaller than for T > T$_v$ (only 2–3 Fe^{2+} ions would contribute to the MR), casting doubts about the suitability of the model in this case. Regarding the magnitude of the MR at maximum field, it presents a maximum at the Verwey transition, corresponding to $\sim$20% (see inset in Fig. 3.13). This peak in T$_v$ has been previously observed in single crystals [28], polycrystalline [29] and epitaxial [15, 30] thin films, in which the existing changes in magnetocrystalline anisotropy play an important role [30]. By measurements at the CCR system we obtain this peak in the three geometries measured, with a value at 11 kOe around 2–4% for 40 and 20 nm thick films (not shown here). These results provide a wide experimental picture of the MR in epitaxial (001) Fe$_3$O$_4$ thin films, where the extrinsic effect of APBs and intrinsic phenomena intermix, disclosing an overall temperature-dependent MR, which does not saturate up to 30 T. A full understanding of this complex scenario is beyond the scope of this study.

3.6 Planar Hall Effect

3.6.1 Introduction to the Planner Hall Effect

In general, in a magnetic material where a current density (**J**) is flowing, an electric field (**E**) is established, proportional to J by means of the resistivity tensor (ρ) [31]:

$$\vec{E} = \tilde{\rho}\vec{J} = \sum_j \rho_{ij}J_j \tag{3.4a}$$

$$\vec{E} = \rho_\perp(B)\vec{J} + \left[\rho_\parallel(B) - \rho_\perp(B)\right](\vec{m}\cdot\vec{J})\vec{m} + \rho_H(B)\vec{m}x\vec{J} \tag{3.4b}$$

where **m** is the normalized magnetization, $\mathbf{m} = \mathbf{M}/M_0$. The second term takes account of the AMR and the PHE, and the third term is the anomalous Hall effect.

In the geometry of measurement of the PHE, both the magnetic field and the current are applied in the film plane. This is in contrast to the standard Hall effect, where the magnetic field is applied perpendicular to the film plane, and the physical origin is different (see Sect. 3.7). In a simple but common situation with a single-domain sample showing in-plane magnetization (**M**), when a current density **J** is applied along the x direction, and **M** forms an angle θ with the x direction, an electric field appears given by

$$E_x = J\rho_\perp + J(\rho_\parallel - \rho_\perp)\cos^2\theta \tag{3.5}$$

$$E_y = J(\rho_\parallel - \rho_\perp)\sin\theta\cos\theta \tag{3.6}$$

The longitudinal component of this electrical field gives rise to the AMR effect, with resistivity $\rho_{xx} = E_x/J$, whereas its transversal component gives rise to the PHE effect, $\rho_{xy} = E_y/J$. Thus, the PHE is a different manifestation of the same physical phenomenon that originates the AMR effect, explained in Sect. 3.5. However, instead of resulting in a change of the longitudinal voltage measured, when changing the angle between the magnetic field and the current, the PHE is a transversal (to the current) voltage. In Fig. 3.14 a scheme of configurations for AMR and PHE measurements is shown.

Due to the link between the magnetization direction and the PHE, the PHE has been used in thin films as a sensitive tool to study in plane magnetization processes when transport measurements are more suitable than direct measurements of the magnetization [32, 33]. The PHE can also be used for sensing low magnetic fields, showing advantages over AMR sensors in terms of thermal drift [34, 35]. Recently, the discovery of the so-called giant planar Hall effect (GPHE), found in the ferromagnetic semiconductor $(Ga_{1-x}Mn_x)As$ [36], has attracted much interest. The effect at liquid helium temperature for this compound is four orders of magnitude higher than that existing in FM metals, reaching values for ρ_{xy} of a few mΩ cm (in FM metals it is of the order of nΩ cm). The discovery has resulted in a thorough study of the magnetization properties of the material by means of this

Fig. 3.14 Geometries for AMR $\left(\rho_{xx} = \frac{\Delta V_x}{I_x}\frac{wt}{L}\right)$ and PHE $\left(\rho_{xy} = \frac{\Delta V_y}{I_x}t\right)$ measurements

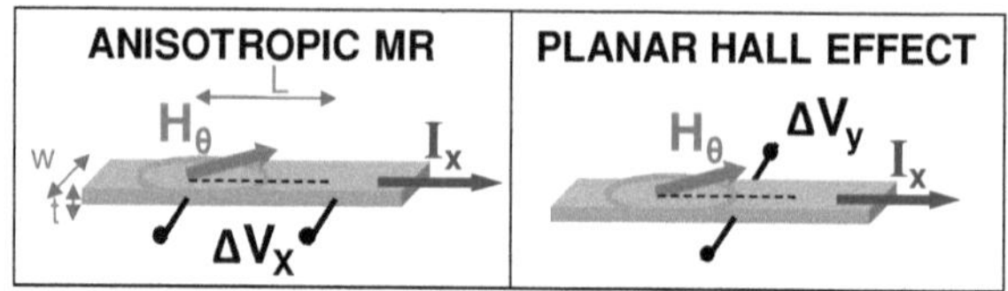

effect [37–39], and as a tool to investigate the resistivity of an individual domain wall [40, 41]. The system (Ga$_{1-x}$Mn$_x$)As, with T$_C$ below 160 K, is restricted for applications of the GPHE in magnetic sensing at low temperatures. In the search for giant PHE at room temperature, promising results have been found in Fe$_3$O$_4$ [42, 43] and manganites [44, 45]. In these films, Bason et al. have even proposed the use of PHE-based MRAM devices [43–45] as an alternative to the more-established MTJ-based technology.

3.6.2 PHE as a Function of Film Thickness (Room Temperature)

At room temperature the PHE has been studied in films with thicknesses of 5, 9, 15, 40 and 150 nm. For the study of the PHE, all measurements were done with the applied magnetic field forming a fixed angle with the current, $\theta = 45°$, since the signal will be maximum in this geometry when saturation in magnetization is reached. From Eq. 3.6:

$$E_y^{(45°)} = \rho_{xy}^{(45°)} J = \frac{\rho_\parallel - \rho_\perp}{2}J \tag{3.7}$$

(For simplicity in notation, we will refer to $\rho_{xy}^{45°}$ as ρ_{xy} from now on.) From these measurements, the AMR can be directly derived:

$$AMR(\%) = 200\frac{\rho_{xy}}{\rho_{xx}} \tag{3.8}$$

To confirm that the even response measured at 45° was caused by the PHE, we also measured with a 135° configuration ($\rho_{xy}^{135°} = -\rho_{xy}^{45°}$), see Eq. 3.6). After subtracting the common offset, signals with opposite sign were obtained (see the particular case for the 40 nm-thick film at room temperature in the Fig. 3.15a). In Fig. 3.15b the results of the PHE for the 40 nm-thick film with current applied in two different directions, **J** // [110] and **J** // [100], are compared (Van Der Pauw and optically lithographed bar, respectively). The sign is different, which indicates that the AMR is positive ($\rho_\parallel > \rho_\perp$) if **J** // [100] whereas it is negative ($\rho_\parallel < \rho_\perp$) if **J** // [110]. Such a difference in the sign of the AMR in Fe$_3$O$_4$ has been explained by Ziese and Blythe with a phenomenological model in which the AMR is expressed as a function of the magnetocrystalline anisotropy constants [16]. As discussed in

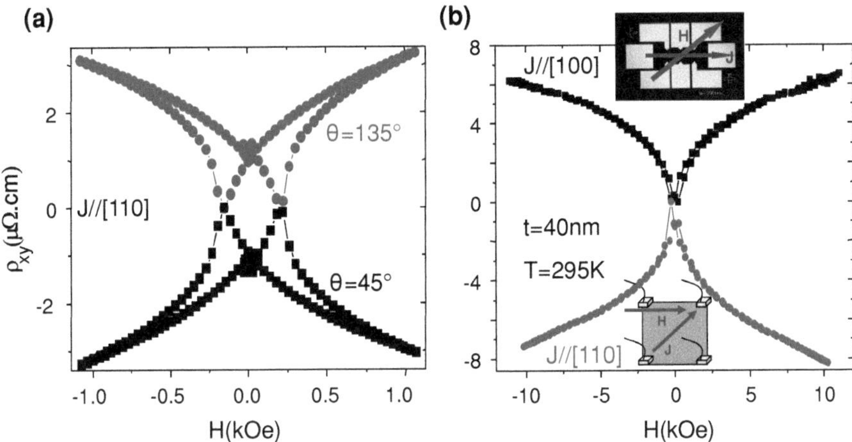

Fig. 3.15 Transversal resistivity as a function of the magnetic field for a 40 nm-thick film at room temperature. **a** Comparison between measurements with the in-plane magnetic field applied forming either $\theta = 45°$ or $\theta = 135°$ with current. **J** // [110] in both cases. **b** Comparison for PHE measurements with two different current directions: **J** // [110]-Van Der Pauw, and **J** // [100]-optically lithographed bar. The origin of the differences of sign in **a** and **b** is discussed in the text

this reference, the AMR sign and its absolute value have an intrinsic origin related to the spin–orbit coupling via the magnetocrystalline anisotropy constants.

In Fig. 3.16 we show the transversal resistivity as a function of the applied magnetic field for all the studied thin films with **J** // [110]. The obtained values are of the same order as in Refs. [42, 43] for Fe_3O_4, where 100 and 9 nm films were, respectively, studied. The values for ρ_{xy} at maximum field are listed in Table 3.1, where a continuous decrease in magnitude with increasing film thickness is observed, as is the case for ρ_{xx}. The maximum value of $\rho_{xy} = -59.4$ $\mu\Omega$ cm is obtained for the 5 nm-thick film. A low-field slope of about 200 $\mu\Omega$ cm/T is found in this film, corresponding to a sensitivity of 400 V/AT, a very large value indeed.

An important issue to point out is the evolution of the films towards super-paramagnetic (SP) behavior as they decrease in thickness, due to the reduction of the AF exchange coupling existing between spins surrounding the APBs [24]. As explained before, the density of APBs varies as $t^{-1/2}$ [10, 11]. The APDs for ultrathin films have sizes of a few unit cells, and their magnetic moments start to fluctuate due to the thermal energy. The progressive shift towards zero field found for the peaks associated with the coercive field in our PHE measurements, as well as the increase of the slope for ρ_{xy} at high fields suggest the occurrence of this phenomenon. These features influence the possible applications of the PHE observed in these films. The SP tendency is also reflected when the AMR is calculated from PHE measurements (see Table 3.1). The AMR has an approximately constant value of -0.2% for films above 10 nm, in agreement with bulk results and thin films in this range of thickness for **J** // [110] [16]. Below 10 nm, an important decrease in magnitude occurs, compatible with the absence of

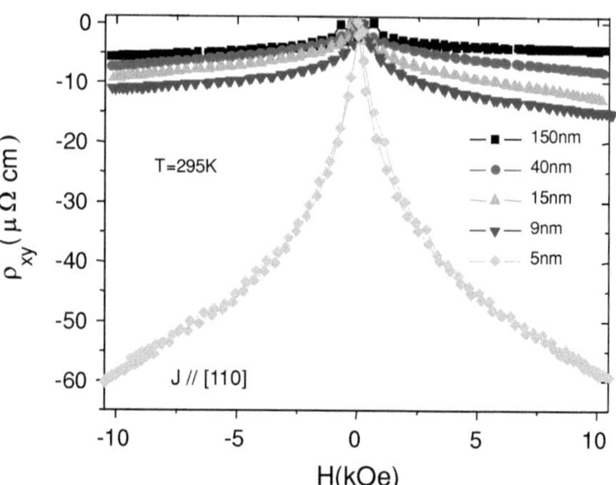

Fig. 3.16 Transversal resistivity ρ_{xy} as a function of the magnetic field ($\theta = 45°$) for several thin film thicknesses at room temperature with current direction **J** // [110]

Table 3.1 Longitudinal (ρ_{xx}) and transversal (ρ_{xy}) resistivity (at maximum field) for the studied thin films at room temperature

Thickness (nm)	150	40	15	9	5
ρ_{xx} (mΩ cm)	5.5	7.0	12.4	23.4	121.4
ρ_{xy} at 11 kOe ($\mu\Omega$ cm)	−4.5	−8.3	−12.9	−15.3	−59.4
AMR (%) = $200\rho_{xy}/\rho_{xx}$	−0.18	−0.24	−0.21	−0.13	−0.1

Measurements were done with **J** applied in the [110] direction. The values for the transversal resistivity are obtained with $\theta = 45°$

anisotropy found for SP samples [24]. This behavior is the same as MR measurements shown in Sect. 3.5.2. Thus, for very thin films (below 5 nm) the PHE will be limited in magnitude by the difficulty in magnetizing the sample and the decrease in the AMR value. Therefore, thinner films than those studied here are not expected to present higher PHE values than the reported ones. The difficulty of using the GPHE in very thin Fe$_3$O$_4$ films for magnetic storage should also be noted, since an improvement in signal magnitude with decreasing thickness also implies a decrease in coercive field. On the other hand, the SP tendency implies a less-hysteretical GPHE effect, which is welcome for magnetic sensing.

3.6.3 PHE as a Function of Temperature

In Fig. 3.17 we show the ρ_{xy} isotherms from room temperature down to 70 K for a 20 nm-thick film. The longitudinal offset makes measurements at lower temperatures unfeasible. The ρ_{xy} increases moderately when cooling from room temperature down to 200 K, changing its sign at about 150 K. When approaching the Verwey transition, a huge increment in the magnitude of ρ_{xy} occurs, as is also observed in the longitudinal resistivity. *Colossal* values, one order of magnitude

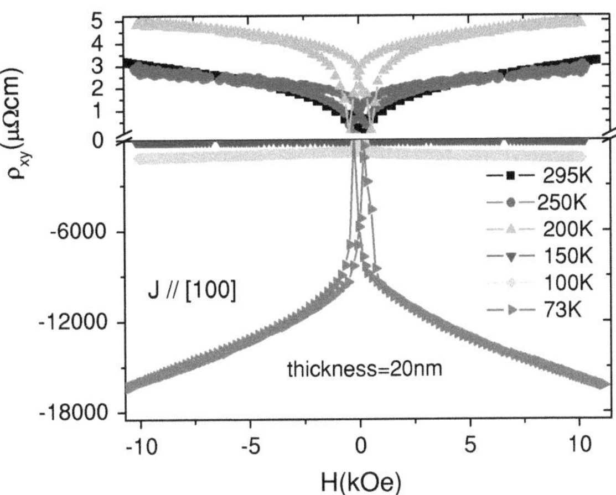

Fig. 3.17 Transversal resistivity isotherms as a function of the magnetic field ($\theta = 45°$) for a 20 nm-thick film. A change in sign for $T < 150$ K is observed. Colossal values are found for $T < T_v$. Measurements done with **J** // [100]

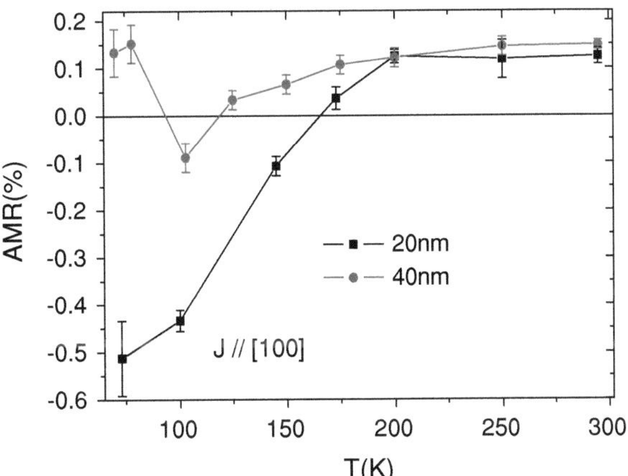

Fig. 3.18 AMR ratio as a function of temperature for a 20 nm and a 40 nm-thick film at 11 kOe. The AMR ratio is calculated from PHE measurements at 45° (Eq. 3.8). Measurements done with **J** // [100]

bigger than the highest reported previously at 4.2 K in $(Ga_{1-x}Mn_x)As$ [42] are found for $T < T_v$. Thus, ρ_{xy} is about 16 mΩ cm at $T = 73$ K in this film.

In the case of a 40 nm-thick film, two changes in sign are observed between room temperature and 70 K. A huge increment in the absolute value of ρ_{xy} is also observed when crossing T_v, giving as a result ρ_{xy} about 15 mΩ cm at $T = 70$ K. In order to gain more insight into the mechanisms responsible for the dependence with temperature of the PHE, AMR ratios calculated from the PHE at 11 kOe in both samples are shown in Fig. 3.18. The obtained values are in agreement with previous studies for the AMR in magnetite thin films [16], with **J** // [100].

For both films, AMR $\approx 0.2\%$, independent of temperature, is found for temperatures above 150 K. Different tendencies are followed in each thin film below this temperature. Whereas in the case of the 20 nm-thick film, a change in sign occurs, giving as a result negative values for $T < 150$ K, for the 40 nm film ρ_{xy} is positive except in the surroundings of T_v. The changes of sign in the AMR are likely to be related to intrinsic changes in the magnetocrystalline anisotropy in this temperature range as previously observed in bulk single crystals [46]. Such changes will dramatically modify the anisotropy constants, which correspondingly will vary the AMR/PHE value as demonstrated in the phenomenological model presented in Ref. [16]. The magnetic anisotropy in these films is expected to depend strongly on the film thickness and detailed magnetic studies beyond the scope of the present study would be required to clarify the quantitative relationship between the magnetic anisotropy changes and the AMR/PHE.

One important result inferred from the obtained AMR temperature dependence is that the absolute value of the AMR is large below the Verwey transition. The ground state of Fe$_3$O$_4$ below the Verwey transition is still a matter of debate [47]. Recent experiments indicate that pure electrostatic models do not seem to be able to explain the ground state below the Verwey transition and the electron-lattice coupling must play a key role in order to stabilize it [47]. The large values of the AMR below the Verwey transition observed by us give evidence of a substantial magnetocrystalline anisotropy of the ground state caused by a significant spin–orbital coupling. This suggests that the claimed strong electron-lattice coupling could arise from a large spin–orbital coupling.

3.7 Anomalous Hall Effect

3.7.1 Introduction to the Anomalous Hall Effect

In non-magnetic materials, when a magnetic field is applied perpendicular to the film plane (H$_z$), a transversal voltage (V$_y$) is established when a current is injected in the x direction. This effect, called the ordinary Hall effect (OHE) is proportional to the applied magnetic field, and is originated by the Lorentz force acting on moving charged carriers, deflecting their trajectory, and establishing a stationary voltage (see Fig. 3.19).

In magnetic materials, an additional term to the OHE, the so called anomalous or extraordinary Hall effect (AHE) appears, as a consequence of the spontaneous magnetization the material possesses. The last term in Eq. 3.4b takes account of the AHE. Thus, the Hall (transversal) resistivity can be written as

$$\rho_{xy} = \rho_H = \mu_0(R_0 H + R_A M) \tag{3.9}$$

Fig. 3.19 Geometry for Hall effect measurements $\left(\rho_{xy} = \frac{\Delta V_y}{I_x}t\right)$. See text for details

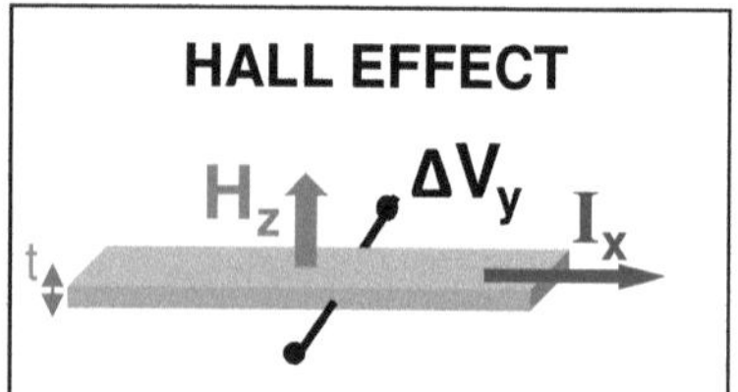

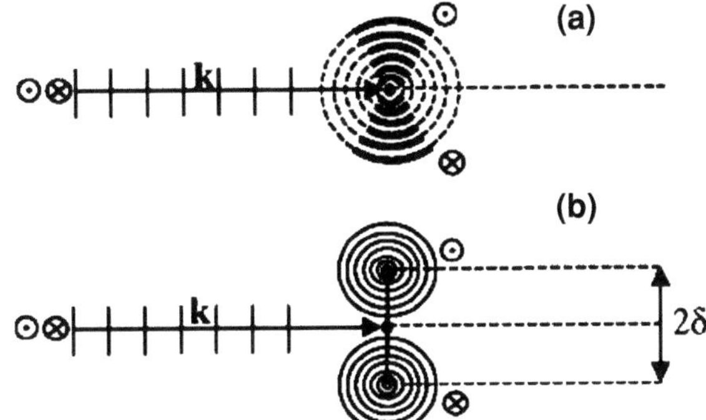

Fig. 3.20 Schematic picture of the two extrinsic mechanisms for the AHE. **a** Skew scattering. **b** Side-jump scattering. The point and cross circles symbols denote up and down spins, respectively. Taken form Crépieux et al. [51]

The first term, proportional to the applied magnetic field, is the OHE, and the second one, in general much larger than the first, is the AHE, which is proportional to the magnetization of the material. The origin of the AHE has been a controversial issue for decades, and various theories have tried to explain this effect. Different dependences between the Hall and the longitudinal resistivity, ρ_{xx}, are given, depending on which origin is associated:

- Extrinsic mechanism, based on the asymmetric interaction of carriers with the scattering centers in the system (see Fig. 3.20). The asymmetries require spin–orbit coupling and spin polarization of carriers to occur. There are two types of mechanisms proposed:

  - *Skew scattering* [48, 49]: Asymmetric scattering, with $\rho_H \propto \rho_{xx}$ ($\sigma_H \propto \sigma_{xx}$).
  - *Side-jump scattering* [50]: Spin-dependent lateral displacement (~ 0.1 Å) of the incident electron wave during the scattering. $\rho_H \propto \rho_{xx}^2$ (σ_H constant).

- Intrinsic mechanism, independent of the scattering of electrons, based on the *Berry phase* of Bloch waves [52, 53]. An additional term to the group velocity of electrons appears, mediated by the Berry phase $\Omega(\mathbf{k})$:

$$v_g(\mathbf{k}) = \frac{1}{\hbar}\nabla_k \varepsilon(\mathbf{k}) - e\mathbf{E} \times \Omega(\mathbf{k}) \tag{3.10}$$

The AHE, via the spin–orbit interaction, can be expressed in terms of this geometrical curvature in momentum space, giving a dependence $\rho_H \propto \rho_{xx}^2$ (σ_H constant).

A recent theory [54] based on multi-band ferromagnetic metals with dilute impurities seems to have solved this complex scenario, where three regimes can be distinguished as a function of the longitudinal conductivity (see Fig. 3.21):

- In the clean regime, with extremely high conductivity, the skew scattering causes the effect ($\sigma_H \propto \sigma_{xx}$).
- An extrinsic-to-intrinsic crossover occurs at lower conductivities ($10^4 \, \Omega^{-1} \, \text{cm}^{-1} < \sigma_{xx} < 10^6 \, \Omega^{-1} \, \text{cm}^{-1}$), where σ_H becomes constant.

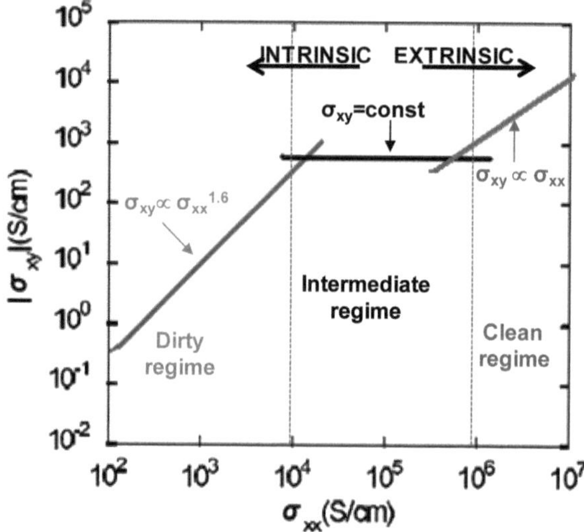

Fig. 3.21 Schematic diagram for the three AHE regimes, classified as a function of the longitudinal conductivity: *Clean* ($\sigma_{xx} > 10^6 \ \Omega^{-1} \ \text{cm}^{-1}$), dominated by the skew scattering: $\sigma_{xy} \propto \sigma_{xx}$, *Intermediate* ($10^4 < \sigma_{xx} < 10^6 \ \Omega^{-1} \ \text{cm}^{-1}$), dominated by an intrinsic mechanism: constant σ_{xy}, and *dirty*, where the damping of the intrinsic contribution results in $\sigma_{xy} \propto \sigma_{xx}^{1.6}$ [Hall conductivity: $\sigma_H = -\rho_H/(\rho_{xx}^2 + \rho_H^2)$; longitudinal conductivity: $\sigma_{xx} = \rho_{xx}/(\rho_{xx}^2 + \rho_H^2)$]

– In the dirty regime ($\sigma_{xx} < 10^4 \ \Omega^{-1} \ \text{cm}^{-1}$) a relation $\sigma_H \propto \sigma_{xx}^{1.6}$ is predicted, caused by the damping of the intrinsic contribution.

Recently, this crossover has been experimentally found for a series of itinerant ferromagnets [55] and a compilation of an appreciable amount of low-conductivity compounds reveals the expected dependence in the dirty limit, regardless of hopping or metallic conduction [56].

3.7.2 AHE as a Function of the Film Thickness (Room Temperature)

In Fig. 3.22 we represent the Hall resistivity ρ_H at room temperature as a function of the film thickness. For all samples, ρ_H is negative, increasing in modulus as the films become thinner. The value for the 40 nm-thick film at 11 kOe is 13.9 $\mu\Omega$ cm, in excellent agreement with that obtained in Ref. [9] at the same field (13.75 $\mu\Omega$ cm), as expected for similar high-quality films. The maximum Hall slope is found for the 5 nm-thick film, reaching 125 $\mu\Omega$ cm/T, which corresponds to a sensitivity of 250 V/AT. This sensitivity at room temperature is only higher for ferromagnetic materials such as Co$_x$Fe$_{1-x}$/Pt multilayers [13]. It is important to remark that in the range of magnetic fields measured with the system available in our laboratory (H$_{\text{max}}$ = 11 kOe), the contribution of the AHE dominates completely the Hall effect, it being unrealistic to determine the ordinary part from the Hall slope at the

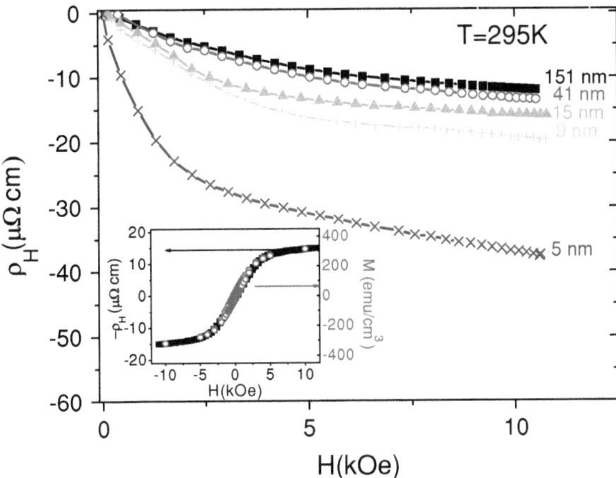

Fig. 3.22 Hall resistivity at room temperature as a function of thickness for a series of samples. The field is applied perpendicular to the plane of the film. The *inset* shows a comparison of the Hall resistivity for a 40 nm thick-film together with the perpendicular magnetization at that temperature for another sample of the same thickness grown in the same conditions. The perfect proportionality between both evidences that the AHE dominates the measurement in the range of field measured

highest measured magnetic field. This impossibility to saturate the film in moderate magnetic fields is due to the antiferromagnetic coupling between spins at the neighborhoods of APBs [17]. This fact can explain the dispersion in values for the ordinary Hall constant R_O obtained by different authors [9, 57]. Therefore, there should exist a perfect proportionality between ρ_H and the magnetization perpendicular to the plane of the film. This is clearly seen, e.g. for the 40 nm-thick film at room temperature in the inset of Fig. 3.23. To reach magnetic saturation, the Hall effect has been studied under very high static magnetic fields(Sect. 3.8).

3.7.3 AHE as a Function of the Temperature

In order to account for the thermal dependence of ρ_H we have measured the 20 and 40 nm films as a function of the temperature. In Fig. 3.23 we show the ρ_H isotherms for the 40 nm-thick film at temperatures above the Verwey transition. For both samples ρ_H increases in modulus monotonously when the temperature diminishes, presenting a huge enhancement as the temperature approaches the transition (see Fig. 3.24). This abrupt change in ρ_H at T_V is a general behavior for Fe_3O_4, also found in single crystals [58] and polycrystalline films [29], which seems to be related with changes in the electronic structure and the spin–orbit coupling that occurs at the transition [3, 4, 59]. In both samples the values for ρ_H increase by a factor higher than 100 from room temperature down to 60 K, reaching values above 1 mΩ cm at that temperature.

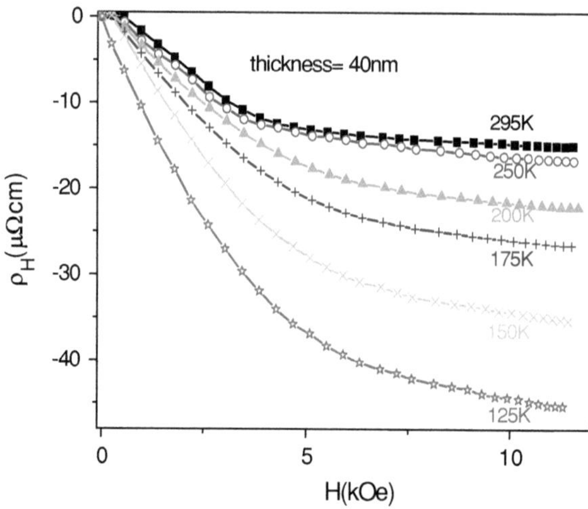

Fig. 3.23 Hall resistivity isotherms in a film of 40 nm for temperatures above the transition. The Hall resistivity increases in modulus as the temperature is diminished, reaching in this case 45 μΩ cm at 125 K

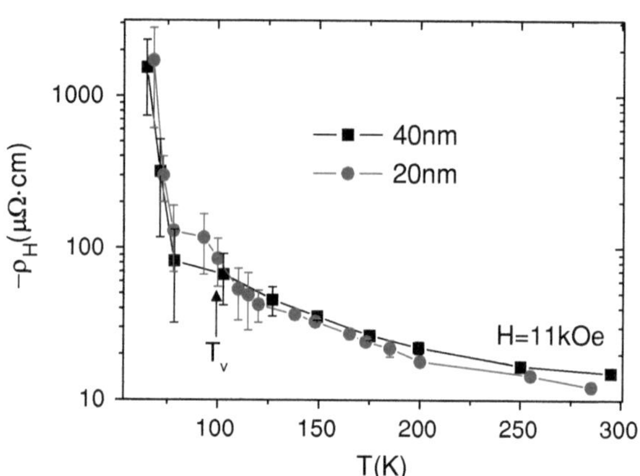

Fig. 3.24 Hall resistivity at 11 kOe as a function of temperature for both samples

3.7.4 AHE in Fe$_3$O$_4$: Universal Behavior

In Fig. 3.25 we show the absolute value of the Hall conductivity at maximum field (11 kOe) as a function of the longitudinal conductivity for all the samples measured, at room temperature and as a function of temperature. In spite of the different thicknesses and measurement temperatures, the relation $\sigma_H \approx 10^{-4}\,\sigma_{xx}^{1.6}$ (dashed line) is followed in all cases. The increase in the value of the error bars for σ_H at temperatures below the transition ($\sigma_{xx} < 1.5\ \Omega^{-1}\,cm^{-1}$) is due to the fact that residual contributions to the voltage measured, such as the longitudinal resistivity, increase substantially. Another relevant source of errors due to slight unavoidable misalignments in the experimental setup is the contribution of the PHE, that, as we have shown in Sect. 3.6, reaches colossal values in our films for $T < T_v$, and could be the origin of previous results yielding a different

Fig. 3.25 Relationship between the magnitude of the Hall conductivity at 11 kOe and the longitudinal conductivity for our films and data taken from Refs. [9, 29, 58, 61]. *Dashed line* is the function f(σ_H) = $10^{-4}\,\sigma_{xx}^{1.6}$

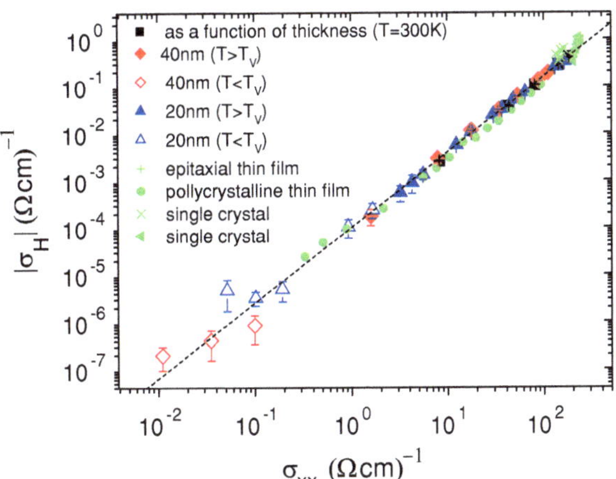

sign of ρ_H [60] in this temperature range. Together with our results, other measurements found in the literature for the AHE in magnetite have also been plotted. Specifically, another 45 nm-thick film epitaxial thin film at room temperature [9], a polycrystalline 250 nm-thick film at temperatures below 300 K [29] and single crystals [58, 61] in the range 150 K < T < 500 K have been included. These data approximately converge with ours, revealing the same physical origin in all cases.

We explained in Sect. 3.1.2 some effects appearing in epitaxial Fe_3O_4 thin films, associated with the presence of APBs. These effects become more evident the thinner the films are. In previous sections we have shown how the APBs influence the magnetotransport properties measured: resistivity, MR and PHE. However, in this section, we have shown that the AHE in Fe_3O_4 thin films presents a universal behavior. First, the enhancement in the Hall resistivity as the temperature is diminished seems to be related with the increase in the longitudinal resistance, as a consequence of the broadening that occurs in the band gap as T_V is approached. This results in huge Hall resistivities of the order of mΩ cm at low temperatures. Second, the dependence $\sigma_H \propto \sigma_{xx}^{1.6}$ found over four decades of longitudinal conductivity is in accordance with the recent unifying theory for the AHE presented in Sect. 3.7.1 . This relation is followed by samples in a wide range of thicknesses and irrespective of the measurement temperature. Furthermore, the same dependence for σ_H is found in other samples of magnetite taken from the literature, regardless of whether they are single crystals or polycrystalline thin films. The theory [54] explains that the magnitude of the AHE is determined by the degree of resonance caused by the location of the Fermi level around an anticrossing of band dispersions. Besides, it assumes metallic conduction. From our results we can infer that the Verwey transition, associated with a cubic-to-monoclinic structure transition, does not play an important role in this behavior. It was pointed out before that this dependence is also followed in hopping conduction [56]. In the case of Fe_3O_4, our results suggest that this dependence also takes place irrespective of the different conduction mechanisms occurring above and below T_V [3]. Another important issue is related with the presence of APBs. As mentioned before, some magnitudes such as the resistivity or magnetization are subjected to important changes from the bulk material to the thin films, due to the presence of these structural defects. The moderate MR ratios in films are explained by the spin-polarized transport through them. However, the AHE is again not affected by the density of APBs, which decreases with film thickness. Finally, it should also be remarked that the presence of stress in epitaxial thin films caused by the substrate could eventually have an influence in the band structure in comparison with the bulk material, but the convergence of all data reveals that this is not an important issue for the AHE. All these facts support the belief that the relation $\sigma_H \propto \sigma_{xx}^{1.6}$ is universal for this low-conductivity regime.

3.7.5 Fe$_3$O$_4$ Inside the AHE Dirty Regime of Conductivities

Finally, for completeness, we show in Fig. 3.26 a broad compilation of results for the AHE found in the literature for magnetic compounds, together with those for Fe$_3$O$_4$ previously shown in Fig. 3.25, all of them in the dirty regime of conduction. We include data for metals, oxides, and magnetically doped semiconductors, with completely different crystallographic structure, magnetic density when doping, and electrical conduction. The Hall conductivity follows the universal relationship $\sigma_H \propto \sigma_{xx}^{1.6}$ over six decades of longitudinal conductivity in all cases. The different proportionality constants in the fits are explained in theory to be caused by a different impurity potential [54]. The universal scaling is followed in spite of the big differences between these compounds. The theory developed in Ref. [54] was for metallic conduction, with the band playing a vital role in this effect, so the good convergence of all these results is surprising.

3.8 Ordinary Hall Effect

In the course of the investigation of the HE using magnetic fields up to 9 T, a constant value of the derivative (which would imply a linear regime for $\rho_H(H)$) is not reached. This implies the lack of magnetic saturation due to the AF coupling between spins surrounding the APBs (Sect. 3.1.2), impeding a clear separation of the OHE from the AHE contribution (Eq. 3.9). In order to reach magnetic saturation, we studied the HE with very high static magnetic fields at the HFML in Nijmegen, The Netherlands (Sect. 2.6) These experiments should permit an accurate determination of the carrier density from the ordinary contribution. From the high-field slope we obtain the ordinary Hall constant:

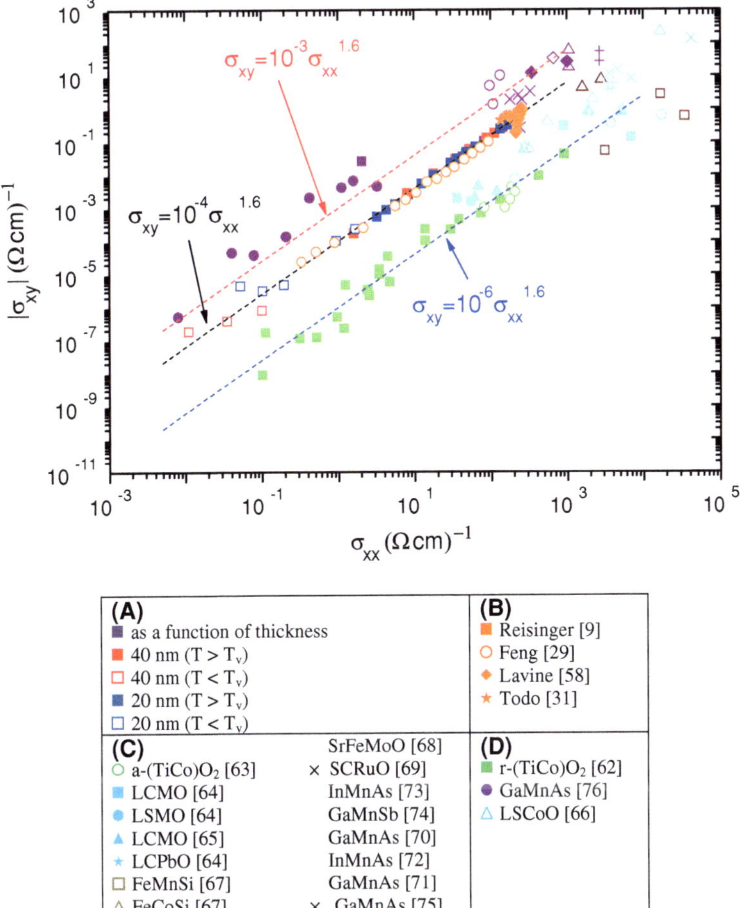

(A)		(B)	
■ as a function of thickness		■ Reisinger [9]	
■ 40 nm (T > T_v)		○ Feng [29]	
□ 40 nm (T < T_v)		◆ Lavine [58]	
■ 20 nm (T > T_v)		★ Todo [31]	
□ 20 nm (T < T_v)			
(C)	SrFeMoO [68]	**(D)**	
○ a-(TiCo)O$_2$ [63]	× SCRuO [69]	■ r-(TiCo)O$_2$ [62]	
■ LCMO [64]	InMnAs [73]	● GaMnAs [76]	
● LSMO [64]	GaMnSb [74]	△ LSCoO [66]	
▲ LCMO [65]	GaMnAs [70]		
★ LCPbO [64]	InMnAs [72]		
□ FeMnSi [67]	GaMnAs [71]		
△ FeCoSi [67]	× GaMnAs [75]		

Fig. 3.26 Anomalous Hall conductivity as a function of the longitudinal conductivity for a broad amount of magnetic oxides, metals and semiconductor ferromagnets, all in the dirty regime of conductivities. All data are well fitted to $\sigma_H \propto \sigma_{xx}^{1.6}$. Most of the data were previously compiled in Ref. [56]. *Below the artwork*, the compounds are shown, together with their references. (A): Our measurements. (B): Other Fe_3O_4 results. (C): Compounds with metallic conduction. (D): Compounds with hopping conduction

$$R_0 = \frac{d\rho_H}{d(\mu_0 H)} \tag{3.11}$$

which gives a density of conducting carriers (in the case of electrons):

$$N = -\frac{1}{R_0 e} \tag{3.12}$$

The transport measurements were carried out by standard AC mode for temperatures above the Verwey transition in static high magnetic fields up to 30 T.

3.8.1 OHE as a Function of the Film Thickness (Room Temperature)

In Fig. 3.27 we show ρ_H up to 30 T for three films selected (thicknesses = 41, 27 and 16 nm) at room temperature. Let us focus first our discussion on the thickest film, t = 41 nm. The dependence of ρ_H(H) at low fields perfectly reproduces the measurements shown in Sect. 3.7.2, and that in reference [9]. In this field range the signal is proportional to M. From approximately 20 T up to 30 T, R_0^{eff} = constant $\approx$ −0.05 $\mu\Omega$ cm/T. This corresponds to an effective electron density $N^{eff} = 1.25 \times 10^{22}$ cm^{-3}, i.e. to n^{eff} = 0.93 electrons per f.u (volume of f.u. = 1/ 8 × volume unit cell = 7.41 × 10^{-23} cm^3). This result is in agreement with the value found by Feng et al. [29] in polycrystalline Fe$_3$O$_4$ thin films, and by Siemons [77] in a single crystal, although in this case he reported a positive R_0 (conduction by holes). From our measurements, $n^{eff} \approx$ 1 e$^-$/f.u. is responsible for the conduction in this film (we understand a slightly smaller value because the sample is mostly, but not completely magnetically saturated).

This value contradicts the results by Lavine et al. [57] in a single crystal, and by Reisinger et al. [9] in a 40–50 nm-thick epitaxial film, of ∼0.25 e$^-$ per f.u. In this last case, they measured with H up to 7 T. If we derive from our measurements the value of R_0 from the slope at that particular range of fields, we find the same result as them (R_0 = −0.2 $\mu\Omega$ cm/T, i.e. 0.22 e$^-$/f.u.). Margulies et al. [17] already demonstrated that thin films are not well saturated up to 9 T due to the AF interactions in APBs, so it is logical to obtain a higher slope, and consequently, a smaller value for n^{eff} than in saturation. $n^{eff} \approx$ 1 e$^-$/f.u. is an experimental

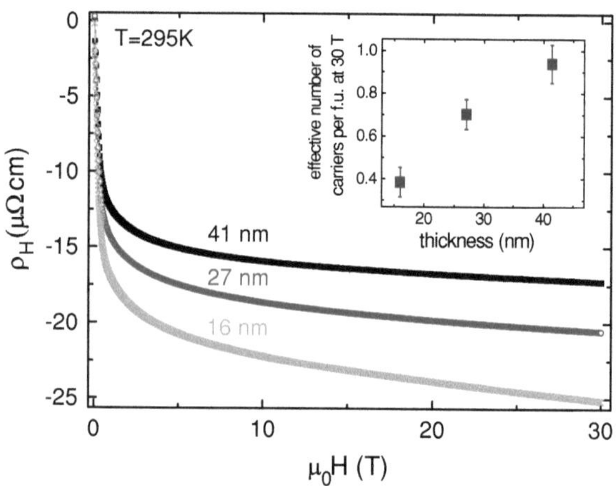

Fig. 3.27 Room temperature Hall resistivity measured up to 30 T, for three samples. In the inset the effective density of electrons at maximum field is shown, where a progressive decrease with thickness is found, as a consequence of a higher density of APBs

Table 3.2 Electrical conduction properties of Fe_3O_4 epitaxial thin films

t (nm)	T (K)	$\sigma\ (\Omega\ cm)^{-1}$	$R_0^{eff}\ (\mu\Omega\ cm/T)$	$N^{eff}\ (e^-/cm^{-3})$	$n^{eff}\ (e^-/f.u.)$	$\mu_H^{eff}\ (cm^2/Vs)$
41	295	129.6	−0.05	1.25×10^{22}	0.93	0.065
27	295	102.2	−0.07	8.93×10^{21}	0.67	0.072
16	295	59.9	−0.12	5.22×10^{21}	0.38	0.074
40*	285	112.6	−0.06	1.08×10^{22}	0.80	0.065
40*	242	88.3	−0.07	9.65×10^{21}	0.71	0.057
40*	150	20.9	−0.11	5.58×10^{21}	0.41	0.023
40*	120	9.0	−0.27	2.29×10^{21}	0.17	0.025

First three rows refer to Van Der Pauw measurements at room temperature. Four last rows are for a 40 nm-thick lithographied sample* as a function of temperature ($T > T_v$). Effective quantities are derived from the slope at an applied field of 30 T

evidence of the appropriate simple picture for the conduction in magnetite at room temperature described by Verwey [2], in terms of thermally activated hopping between B (octahedral) sites (see Sect. 3.1.1). If we do the same treatment for the other two, thinner samples, we find smaller values for n^{eff} (see Table 3.2 and inset of Fig. 3.27 for the exact values). The decrease of n^{eff} with thickness is caused by the APBs, since its density varies as $t^{-1/2}$ [10, 11, 17]. This decrease could be extrinsic, caused by a progressive lack of saturation of films as they get thinner, even at these huge static magnetic fields. However, if we considered that the samples are mostly saturated at 30 T, the progressive decrease of n^{eff} should be intrinsic, due to the electron localization around the increasing density of APBs [10, 11, 16–18]. In all cases the calculated Hall mobility has approximately the same value for the three films, $\mu_H = |R_0|\sigma \approx 0.07\ cm^2/Vs$.

3.8.2 OHE as a Function of Temperature

In Fig. 3.28 the evolution of $\rho_H(H)$ with temperature for a lithographied 40 nm-thick film is shown, at temperatures above the Verwey transition. Apart from the evident increase of the Hall resistivity upon temperature decrease (due to the AHE), as was shown in Sect. 3.7.3 , the high-field slope increases in modulus, becoming more negative. This results in a decrease of the effective number of carriers as the temperature is lowered, with values evolving from $\sim 1\ e^-/f.u.$ at 285 K, down to $\sim 0.2\ e^-/f.u.$ at 120 K, near the temperature where the metal–insulator transition occurs. Above T_v, the Ihle–Lorenz model is normally used to explain the conduction in Fe_3O_4 [3, 5], given by a superposition of small-polaron band, and hopping conductivity. For $T_v < T < 285$ K we can consider an Arrhenius dependence for $n^{eff}(T)$ and $\mu_H(T)$:

$$n^{eff}(T) = n_0 \exp[-E_A(n)/kT] \tag{3.13a}$$

$$\mu_H^{eff}(T) = \mu_H^0 \exp[-E_A(\mu_H)/kT] \tag{3.13b}$$

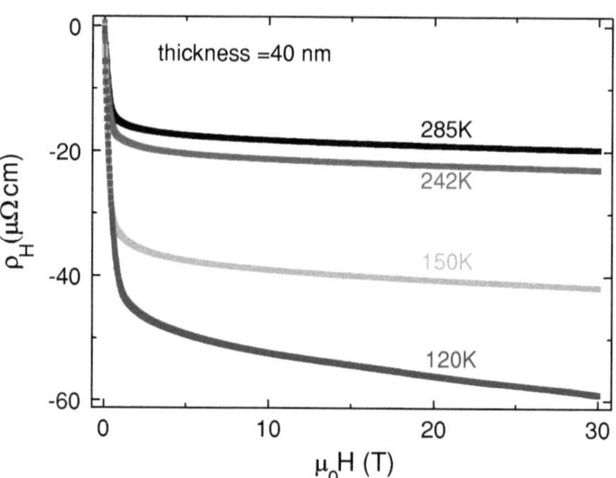

Fig. 3.28 Hall resistivity up to 30 T, for temperatures above the Verwey transition measured in a lithographied 40 nm-thick sample. The ordinary Hall effect increases when temperature is lowered, giving account of the decrease of conducting electrons as the Verwey transition is approached

we obtain $E_A(n) = (26 \pm 4)$ meV and $E_A(\mu_H) = (20 \pm 5)$ meV, these values being in good agreement with those reported in Ref. [29]. The positive temperature dependence for μ_H can be understood by the decrease of the effective mass of the small polarons with temperature. The sum of both activation energies is around 50 meV, a typical reported value for the activation energy of conductivity in this temperature range [3, 29, 57]. The details of the data analysis are also reported in Table 3.2. Unfortunately, we cannot show results for the OHE for temperatures below T_v. Minimum thermal drifts (typical temperature changes during these long experiments were smaller than 1 K) hinder the accurate determination of the Hall slope, due to the large change in magnitude that $\sigma(T)$ and ρ_H present at $T < T_v$ (Sect. 3.7.3).

3.9 Conclusions

We have grown high-quality epitaxial Fe$_3$O$_4$ films on MgO substrates and studied systematically their magnetotransport properties. A complete optical lithography process was successfully performed for the films, allowing technically difficult measurements, below the Verwey transition.

Resistivity measurements show a perfect agreement with previous results of other groups. ρ follows the relation $t^{-1/2}$, as a consequence of the increase of the APBs when the film thickness decreases. Measurements as a function of temperature show a decrease of T_v as films become thinner, together with a smoothing of the Verwey transition, becoming undistinguishable for t < 10 nm.

Magnetoresistance measurements at moderate high fields reproduce previous results in the literature, and can be understood by the spin-polarized transport through APBs. The tendency towards the superparamagnetic behavior of the APDs is evidenced for ultra-thin films. The MR is not saturated at 30 T. A peak for the absolute value of the MR at T_v indicates that an intrinsic phenomenon is super-imposed onto the extrinsic effect caused by APBs. The models previously established in the literature do not give a full explanation of the MR measured.

Thickness and temperature dependence of the giant planar Hall effect in the films have been studied. Samples with moderately high APBs density show higher PHE signals due to the induced increase in the absolute value of the resistivity anisotropy, making them interesting candidates for magnetic sensing and non-volatile memories, as well as for magnetization studies through transport measurements. A record value of the PHE at room temperature of $|\rho_{xy}| \approx 60 \ \mu\Omega$ cm was obtained for the 5 nm-thick film. As a function of temperature, the PHE increases in magnitude, reaching colossal values below the Verwey transition, in the range of a few mΩ cm. The anisotropic MR values inferred from these measurements indicate that the ground state below the Verwey transition is likely to bear a substantial spin–orbit coupling.

Anomalous Hall effect measurements in the films show that the dependence $\sigma_H \propto \sigma_{xx}^{1.6}$ is fulfilled over four decades of longitudinal conductivity, irrespective of thickness or measurement temperature. A comparison with other results in the literature confirms this behavior is general for this compound, as well as for others in the same (low) regime of conduction. This result supports a recent theory developed to explain the AHE, indicating the universality of this relation in the dirty regime of conductivities.

Transport measurements at static magnetic fields of 30 T show that 40 nm-thick films are magnetically saturated, overcoming the strong AF coupling present between APBs. This makes possible to separate the OHE from the anomalous one. We derive that ~ 1 electron per f.u. contributes to the conduction at room temperature. Thinner samples have a smaller effective density of conducting electrons, as a consequence of the APBs. The density of electrons as well as the Hall mobility fit well to an Arrhenius thermal dependence.

References

1. M.L. Glasser, F.J. Milford, Spin wave spectra of magnetite. Phys. Rev. **130**, 1783 (1963)
2. E.J.W. Verwey, Electronic conduction of magnetite (Fe_3O_4) and its transition point at low temperatures. Nature (London) **144**, 327 (1939)
3. F. Walz, The Verwey transition—a topical review. J. Phys.: Condens. Matter **14**, R285 (2002)
4. J. García, G. Subías, The Verwey transition—a new perspective. J. Phys.: Condens. Matter **16**, R145 (2004)
5. D. Ihle, B. Lorenz, Small-polaron band versus hopping conduction in Fe_3O_4. J. Phys. C: Solid State Phys. **18**, L647 (1985)
6. N.F. Mott, *Metal–insulator transitions*, 2nd edn. (Taylor and Francis, London, 1990)
7. A. Yanase, K. Siratori, Band structure in the high temperature phase of Fe_3O_4. J. Phys. Soc. Jpn. **53**, 312 (1984)
8. Z. Zhang, S. Satpathy, Electron states, magnetism, and the Verwey transition in magnetite. Phys. Rev. B **44**, 13319 (1991)
9. D. Reisinger, P. Majewski, M. Opel, L. Alff, R. Gross, Hall effect, magnetization, and conductivity of Fe_3O_4 epitaxial thin films. Appl. Phys. Lett. **85**, 4980 (2004)

10. A.V. Ramos, J.-B. Mouse, M.-J. Guittet, A.M. Bataille, M. Gautier-Soyer, M. Viret, C. Gatel, P. Bayle-Guillemaud, E. Snoeck, Magnetotransport properties of Fe$_3$O$_4$ epitaxial thin films: thickness effects driven by antiphase boundaries. J. Appl. Phys. **100**, 103902 (2006)
11. W. Eerenstein, T.T.M. Palstra, T. Hibma, S. Celotto, Origin of the increased resistivity in epitaxial Fe$_3$O$_4$ films. Phys. Rev. B **66**, 201101(R) (2002)
12. S.K. Arora, R.G.S. Sofin, I.V. Shvets, M. Luysberg, Anomalous strain relaxation behavior of Fe$_3$O$_4$/MgO (100) heteroepitaxial system grown using molecular beam epitaxy. J. Appl. Phys. **100**, 073908 (2006)
13. Y. Zhu, J.W. Cai, Ultrahigh sensitivity Hall effect in magnetic multilayers. Appl. Phys. Lett. **90**, 012104 (2007)
14. J.F. Bobo, D. Basso, E. Snoeck, C. Gatel, D. Hrabovsky, J.L. Gauffier, L. Ressier, R. Mamy, S. Vinsnovsky, J. Hamrle, J. Teillet, A.R. Fert, Magnetic behavior and role of the antiphase boundaries in Fe$_3$O$_4$ epitaxial films sputtered on MgO (001). Eur. Phys. J. B **24**, 43 (2001)
15. G.Q. Gong, A. Gupta, G. Xiao, W. Qian, V.P. Dravid, Magnetoresistance and magnetic properties of epitaxial magnetite thin films. Phys. Rev. B **56**, 5096 (1997)
16. M. Ziese, H.J. Blythe, Magnetoresistance of magnetite. J. Phys. Condens. Matter **12**, 13 (2000)
17. D.T. Margulies, F.T. Parker, M.L. Rudee, F.E. Spada, J.N. Chapman, P.R. Aitchison, A.E. Berkowitz, Origin of the anomalous magnetic behavior in single crystal Fe$_3$O$_4$ films. Phys. Rev. Lett. **79**, 5162 (1997)
18. W. Eerenstein, T.T.M. Palstra, S.S. Saxena, T. Hibma, Spin-polarized transport across sharp antiferromagnetic boundaries. Phys. Rev. Lett. **88**, 247204 (2002)
19. D.J. Huang, C.F. Changa, J. Chena, L.H. Tjengc, A.D. Ratad, W.P. Wua, S.C. Chunga, H.J. Lina, T. Hibmad, C.T. Chen, Spin-resolved photoemission studies of epitaxial Fe$_3$O$_4$(100) thin films. J. Magn. Magn. Mat. **239**, 261 (2002)
20. D. Serrate, J.M. De Teresa, P.A. Algarabel, R. Fernández-Pacheco, M.R. Ibarra, J. Galibert, Grain boundary magnetoresistance up to 42 Tesla in cold-pressed Fe$_3$O$_4$ nanopowders. J. Appl. Phys. **97**, 084322 (2005)
21. L.J. van der Pauw, A method of measuring the resistivity and Hall coefficient on lamellae of arbitrary shape. Philips Tech. Rev. **36**, 220 (1958)
22. X.W. Li, A. Gupta, G. Xiao, G.Q. Gong, Transport and magnetic properties of epitaxial and polycrystalline magnetite thin films. J. Appl. Phys. **83**, 7049 (1998)
23. J.P. Wright, J.P. Attfield, P.G. Radaelli, Long range charge ordering in magnetite below the Verwey transition. Phys. Rev. Lett. **87**, 266401 (2001)
24. W. Eerenstein, T. Hibma, S. Celotto, Mechanism for superparamagnetic behavior in epitaxial Fe$_3$O$_4$ films. Phys. Rev. B **70**, 184404 (2004)
25. J. Kondo, Anomalous Hall effect and magnetoresistance of ferromagnetic metals. Prog. Theor. Phys. **27**, 772 (1962)
26. D. Venkateshvaran, M. Althammer, A. Nielsen, S. Geprägs, M.S. Ramachandra Rao, S.T.B. Goennenwein, M. Opel, R. Gross, Epitaxial Zn$_x$Fe$_{3-x}$O$_4$ thin films: a spintronic material with tunable electrical and magnetic properties. Phys. Rev. B **79**, 134405 (2009)
27. M. Viret, L. Ranno, J.M.D. Coey, Magnetic localization in mixed-valence manganites. Phys. Rev. B **55**, 8067 (1997)
28. V.V. Gridin, G.R. Hearne, J.M. Honing, Magnetoresistance extremum at the first-order Verwey transition in magnetite (Fe$_3$O$_4$). Phys. Rev. B **53**, 15518 (1996)
29. J.S.-Y. Feng, R.D. Pashley, M.-A. Nicolet, Magnetoelectric properties of magnetite thin films. J. Phys. C: Solid State Phys. **8**, 1010 (1975)
30. S.B. Ogale, K. Ghosh, R.P. Sharma, R.L. Greene, R. Ramesh, T. Venkatesan, Magnetotransport anisotropy effects in epitaxial magnetite (Fe$_3$O$_4$) thin films. Phys. Rev. B **57**, 7823 (1998)
31. I.A. Campbell, A. Fert, in *Ferromagnetic Materials*, vol. 3, ed. by E.P. Wohlfarth (North-Holland, Amsterdam, 1982), p. 747
32. G. Li, T. Yang, Q. Hu, H. Jiang, W. Lai, Ferromagnetic domain structure and hysteresis of exchange bias in NiFe/NiMn bilayers. Phys. Rev. B **65**, 134421 (2002)

33. S.T.B. Goennenwein, R.S. Keizer, S.W. Schink, I. van Dijk, T.M. Klapwijk, G.X. Miao, G. Xiao, A. Gupta, Planar Hall effect and magnetic anisotropy in epitaxially strained chromium dioxide thin films. Appl. Phys. Lett. **90**, 142509 (2007)
34. A. Schuhl, F. Nguyen Van Dau, J.R. Childress, Low-field magnetic sensors based on the planar Hall effect. Appl. Phys. Lett. **66**, 2751 (1995)
35. A.O. Adeyeye, M.T. Win, T.A. Tan, G.S. Chong, V. Ng, T.S. Low, Planar Hall effect and magnetoresistance in Co/Cu multilayer films. Sens. Actuators A **116**, 95 (2004)
36. H.X. Tang, R.K. Kawakami, D.D. Awschalom, M.L. Roukes, Giant Planar Hall effect in epitaxial (Ga, Mn)As devices. Phys. Rev. Lett. **90**, 107201 (2003)
37. K.Y. Wang, K.W. Edmonds, R.P. Camion, L.X. Zhao, C.T. Foxon, B.L. Gallagher, Anisotropic magnetoresistance and magnetic anisotropy in high-quality (Ga, Mn)As films. Phys. Rev. B **72**, 085201 (2005)
38. W.L. Lim, X. Liu, K. Dziatkowski, Z. Ge, S. Shen, J.K. Furdyna, M. Dobrowolska, Investigation of magnetocrystalline anisotropy by planar Hall effect in GaMnAs epilayers grown on vicinal GaAs substrates. J. Appl. Phys. **99**, 08D505 (2006)
39. D.Y. Shin, S.J. Chung, S. Lee, X. Liu, J.K. Furdyna, Stable multidomain structures formed in the process of magnetization reversal in GaMnAs ferromagnetic semiconductor thin films. Phys. Rev. Lett. **98**, 047201 (2007)
40. H.X. Tang, S. Masmanidis, R.K. Kawakami, D.D. Awschalom, M.L. Roukes, Negative intrinsic resistivity of an individual domain wall in epitaxial (Ga, Mn)As microdevices. Nature **43**, 52 (2004)
41. H.X. Tang, M.L. Roukes, U.S. Patent 6910382 (2004)
42. X. Jin, R. Ramos, Y. Zhou, C. McEvoy, I.V. Shvets, Planar Hall effect in magnetite (100) films. J. Appl. Phys. **99**, 08C509 (2006)
43. Y. Bason, L. Klein, H.Q. Wang, J. Hoffman, X. Hong, V.E. Henrich, C.H. Ahn, Planar Hall effect in epitaxial thin films of magnetite. J. Appl. Phys. Lett. **101**, 09J507 (2007)
44. Y. Bason, L. Klein, J.-B. Yau, X. Hong, C.H. Ahn, Giant planar Hall effect in colossal magnetoresistive $La_{0.84}Sr_{0.16}MnO_3$ thin films. Appl. Phys. Lett. **84**, 2593 (2006)
45. Y.Y. Bason, L. Klein, J.-B. Yau, X. Hong, J. Hoffman, C.H. Ahn, Planar Hall-effect magnetic random access memory. J. Appl. Phys. **99**, 08R701 (2006)
46. V. Skumryev, H.J. Blythe, J. Cullen, J.M.D. Coey, AC susceptibility of a magnetite crystal. J. Magn. Magn. Mater. **196–197**, 515 (1999)
47. I. Leonov, A.N. Yaresko, On the Verwey charge ordering in magnetite. J. Phys.: Condens. Matter **19**, 021001 (2007)
48. J. Smit, The spontaneous Hall effect in ferromagnetics. Physica (Utretch) **21**, 877 (1955)
49. J. Smit, The spontaneous Hall effect in ferromagnetics II. Physica (Utretch) **24**, 39 (1958)
50. L. Berger, Side-jump mechanism for the Hall effect of ferromagnets. Phys. Rev. B **2**, 4559 (1970)
51. A. Crépieux, P. Bruno, Theory of the anomalous Hall effect from the Kubo formula and the Dirac equation. Phys. Rev. B **64**, 014416 (2001)
52. R. Karplus, J.M. Luttinger, Hall effect in ferromagnetics. Phys. Rev. **95**, 1154 (1954)
53. G. Sundaram, Q. Niu, Wave-packet dynamics in slowly perturbed crystals: gradient corrections and Berry-phase effects. Phys. Rev. B **59**, 14915 (1999)
54. S. Onoda, N. Sugimoto, N. Nagaosa, Intrinsic versus extrinsic anomalous Hall effect in ferromagnets. Phys. Rev. Lett. **97**, 126602 (2006)
55. T. Miyasato, N. Abe, T. Fujii, A. Asamitsu, S. Onoda, Y. Onose, N. Nagaosa, Y. Tokura, Crossover behavior of the anomalous Hall effect and anomalous Nernst effect in itinerant ferromagnets. Phys. Rev. Lett. **99**, 086602 (2007)
56. T. Fukumura, H. Toyosaki, K. Ueno, M. Nakano, T. Yamasaki, M. Kawasaki, A scaling relation of anomalous Hall effect in ferromagnetic semiconductors and metals. Jpn. J. Appl. Phys. **46**, 26 (2007)
57. J. Lavine, Ordinary Hall effect in $(Fe_3O_4)_{0.75}(FeO)_{0.25}(Fe_2O_3)$ at room temperature. Phys. Rev. **114**, 482 (1959)

58. J. Lavine, Extraordinary Hall-effect measurements on Ni, Some Ni Alloys, and ferrites. Phys. Rev. **123**, 1273 (1961)
59. V. Brabers, F. Waltz, H. Kronmuller, Impurity effects upon the Verwey transition in magnetite. Phys. Rev. B **58**, 14163 (1998)
60. D. Kostopoulos, A. Theodossiou, Hall effect in magnetite. Phys. Status Solidi A **2**, 73 (1970)
61. S. Todo, K. Siratori, S. Kimura, Transport properties of the high temperature phase of Fe$_3$O$_4$. J. Phys. Soc. Jpn. **64**, 6 (1995)
62. H. Toyosaki, T. Fukumura, Y. Yamada, K. Nakajima, T. Chikyow, T. Hasegawa, H. Koinuma, M. Kawasaki, Anomalous Hall effect governed by electron doping in a room-temperature transparent ferromagnetic semiconductor. Nat. Mater. **3**, 221 (2004)
63. K. Ueno, T. Fukumura, H. Toyosaki, M. Nakano, M. Kawasaki, Anomalous Hall effect in anatase Ti$_{1-x}$Co$_x$O$_2$-delta at low temperature regime. Appl. Phys. Lett. **90**, 072103 (2007)
64. Y. Lyanda-Geller, S.H. Chun, M.B. Salamon, P.M. Goldbart, P.D. Han, Y. Tomioka, A. Asamitsu, Y. Tokura, Charge transport in manganites: Hopping conduction, the anomalous Hall effect, and universal scaling. Phys. Rev. B **63**, 184426 (2001)
65. P. Matl, N.P. Ong, Y.F. Yan, Y.W. Li, D. Studebaker, T. Baum, G. Doubinina, Hall effect of the colossal magnetoresistance manganite La$_{1-x}$Ca$_x$MnO$_3$. Phys. Rev. B **57**, 10248 (1998)
66. Y. Onose, Y. Tokura, Doping dependence of the anomalous Hall effect in La$_{1-x}$Sr$_x$CoO$_3$. Phys. Rev. B **73**, 174421 (2006)
67. N. Manyala, Y. Sidis, J.F. Ditusa, G. Aeppli, D.P. Young, Z. Fisk, Large anomalous Hall effect in a silicon-based magnetic semiconductor. Nat. Mater. **3**, 255 (2004)
68. Y. Tomioka, T. Okuda, Y. Okimoto, R. Kumai, K.-I. Kobayashi, Y. Tokura, Magnetic and electronic properties of a single crystal of ordered double perovskite Sr$_2$FeMoO$_6$. Phys. Rev. B **61**, 422 (2000)
69. R. Mathieu, A. Asamitsu, H. Yamada, K.S. Takahashi, M. Kawasaki, Z. Fang, N. Nagaosa, Y. Tokura, Scaling of the anomalous Hall effect in Sr$_{1-x}$Ca$_x$RuO$_3$. Phys. Rev. Lett. **93**, 016602 (2004)
70. F. Matsukura, E. Abe, H. Ohno, Magnetotransport properties of (Ga, Mn)Sb. J. Appl. Phys. **87**, 6442 (2000)
71. K.W. Edmonds, R.P. Campion, K.-Y. Wang, A.C. Neumann, B.L. Gallagher, C.T. Foxon, P.C. Main, Magnetoresistance and Hall effect in the ferromagnetic semiconductor Ga$_{1-x}$Mn$_x$As. J. Appl. Phys. **93**, 6787 (2003)
72. A. Oiwa, A. Endo, S. Katsumoto, Y. Iye, H. Ohno, Magnetic and transport properties of the ferromagnetic semiconductor heterostructures (In, Mn)As/(Ga, Al)Sb. Phys. Rev. B **59**, 5826 (1999)
73. H. Ohno, H. Munekata, T. Penny, S. von Molnar, L.L. Chang, Magnetotransport properties of p-type (In, Mn)As diluted magnetic III-V semiconductors. Phys. Rev. Lett. **68**, 2664 (1992)
74. F. Matsukura, H. Ohno, A. Shen, Y. Sugawara, Transport properties and origin of ferromagnetism in (Ga, Mn)As. Phys. Rev. B **57**, R2037 (1998)
75. D. Chiba, Y. Nishitani, F. Matsukura, H. Ohno, Properties of Ga$_{1-x}$Mn$_x$As with high Mn composition (x > 0.1). Appl. Phys. Lett. **90**, 122503 (2007)
76. S.U. Yuldashev, H.C. Jeon, H.S. Im, T.W. Kang, S.H. Lee, J.K. Furdyna, Anomalous Hall effect in insulating Ga$_{1-x}$Mn$_x$As. Phys. Rev. B **70**, 193203 (2004)
77. W.J. Siemons, Hall mobility measurements on magnetite above and below the electronic ordering temperature. IBM J. Res. Dev. **14**, 245 (1970)

Chapter 4
Conduction in Atomic-Sized Magnetic Metallic Constrictions Created by FIB

This chapter reports the study of electron conduction in metallic constrictions, with sizes in the atomic range. First, a novel technique for the fabrication of nano-constrictions at room temperature in metallic materials will be presented, based on the potentialities that the dual-beam equipment offers. We will show that it is possible to create atomic-sized constrictions inside the vacuum chamber where the sample is fabricated. Second, this method has been used for experiments in a magnetic metal such as iron, where phenomena such as ballistic magnetoresistance and ballistic anisotropic magnetoresistance have been investigated.

4.1 Introduction

We described in Sect. 1.6.2 the interesting effects appearing in magnetic atomic-sized constrictions, together with the controversies associated with these high-impact results. Thus, the creation of magnetic nanocontacts free from spurious effects appears to be of considerable importance, in order to clearly discriminate whether these effects are intrinsic or not in magnetic materials, for its possible use in spintronics. The typical techniques used for this purpose: STM, MBJ, EBJ, ECJ (see Sect. 1.6.1) control extraordinarily well the reduction of the contacts to the atomic regime. However, the constrictions fabricated by these methods are not tightly attached to a substrate, favoring the presence of mechanical artifacts. Besides, the construction of real devices using BMR or BAMR with those methods is not possible, due to their incompatibility for integration. Therefore, the use of nanolithography techniques aiming to create stable nanoconstrictions at room temperature, could lead to immense applications.

There exist some examples in the literature using EBL or FIB to create contacts [1–6]. However, the reduction of the nanoconstrictions to effective sizes smaller

A. Fernandez-Pacheco, *Studies of Nanoconstrictions, Nanowires and Fe₃O₄ Thin Films*, Springer Theses, DOI: 10.1007/978-3-642-15801-8_4,

than 100 nm is really challenging. The MR effects found are, in general, similar to those for bulk, and the lack of reproducibility is evident.

The work presented in this chapter also uses the FIB as the nanolithography tool, but has as a novelty that the electrical resistance is controlled at the same time as the contact is decreasing in size. This allows a perfect control of the device while it is approaching to the ballistic regime of conduction. The work has been successful in the realization of an approach that creates atomic-sized contacts, although the study of MR effects in magnetic nanoconstrictions has, on the other hand, been less satisfactory.

4.2 Experimental Procedure. Example for a Non-Magnetic Material: Chromium

For the realization of atomic-sized constrictions we have designed an experimental procedure basically consisting of the following steps (see details of the equipment in Chap. 2):

1. Optical lithography process via lift-off to define a 4 µm metallic path connected by macroscopic pads.
2. FIB etching process in the Dual Beam system, controlling the resistance → creation of metallic constriction.
3. If the process is done with iron, chamber venting and measurement of the MR as a function of temperature of the CCR system.

In order to illustrate the procedure, the creation of a controlled atomic-sized constriction based on chromium/aluminum electrodes is described (see a scheme of the process in Fig. 4.1). The experiments were made at room temperature in the Dual Beam system. Two electrical microprobes were contacted on the pads (see Fig. 4.2). These conductive microprobes were connected via a feedthrough to the Keithley system explained in Sect. 2.2.2), located out of the chamber. The lead resistance is about 15 Ω, which guarantees no significant influence on the measured resistance at the stages where the conductance steps occur, which corresponds to resistances of the order of 10 kΩ.

The Ga-ion etching process was done in two consecutive steps. First, starting from 15 Ω, with the ion-column set at 5 kV and 0.12 nA, aiming to shorten the time of the experiment. The etching was stopped when the resistance reached 500 Ω, still far from the metallic quantum limit. In the second step, the ion-column was set at 5 kV and 2 pA, for about 12 min. During this period we monitored the variation of the resistance and simultaneously 10 kV-SEM images were collected. This allowed the correlation between resistance and micro-structure. In other experiments, the milling was stopped at determined resistance values, at which current versus voltage curves were measured. The maximum bias current was selected in order to avoid the deterioration of the device due to heating effects.

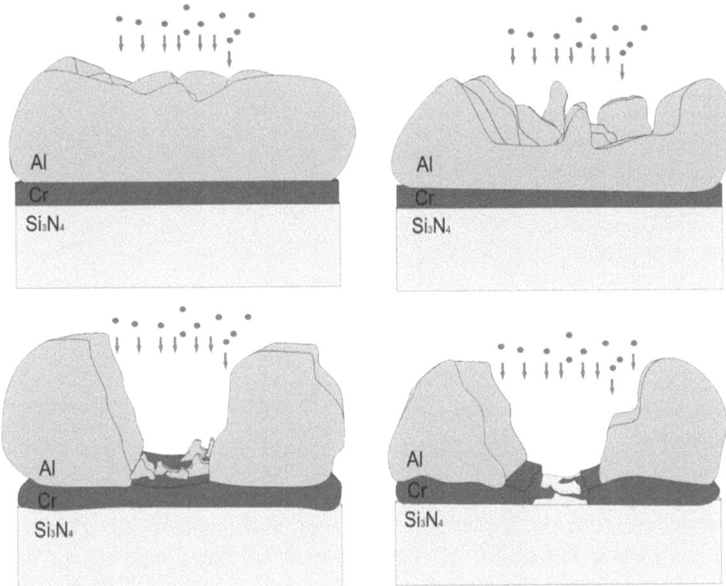

Fig. 4.1 Schematic cartoon of the etching process. Electrical transport measurements are simultaneously measured. The inhomogeneous etching in Al favours clear images of the formed nanoconstrictions in Cr when Al is fully removed

Fig. 4.2 SEM image of the electrical aluminum pads patterned by optical lithography. The two microprobes are contacted for real-time control of the electrode resistance. The *square* indicates the area of etching. The *inset* is an SEM image of the Al/Cr electrode prior to the Ga etching

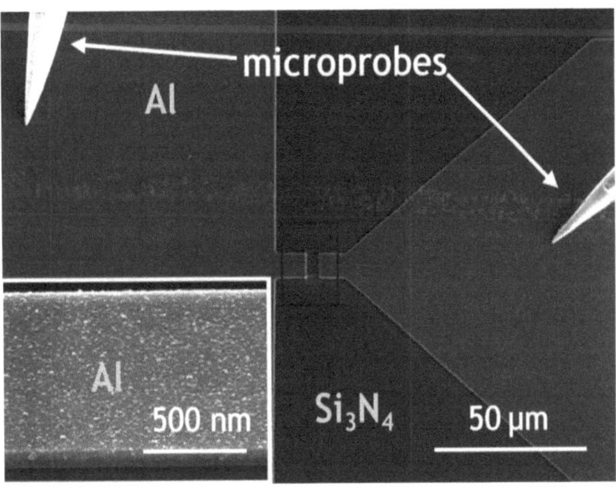

In Fig. 4.3, we show a typical resistance-versus-etching time result at a starting value of 500 Ω. The resistance increases continuously for about 10 min, up to a point in which three orders of magnitude jump is seen. Just before this abrupt change, the resistance shows a discrete number of steps. We can see in the inset that, for this particular case, plateaus with a small negative slope for the

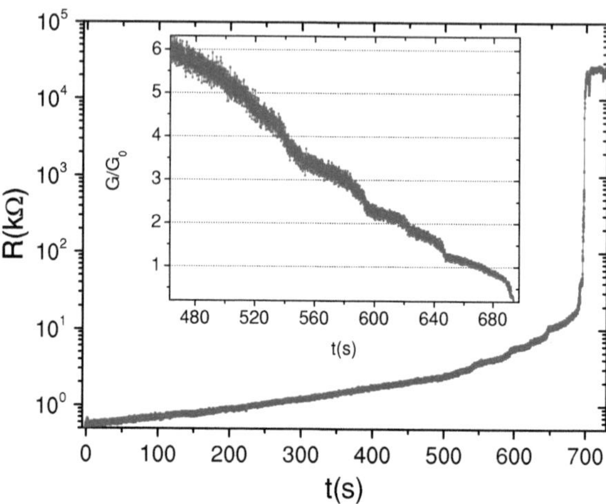

Fig. 4.3 Resistance versus time in a typical ion etching process for Cr/Al. Conductance (in $G_0 = 2e^2/h$ units) versus etching time of the nanoconstriction whose results are shown in the *inset*. In this stage of milling, steps of the order of G_0 are seen, corresponding to discrete thinning of the contact area of the order of one atom. The last step is followed by a sharp decrease of conductance, corresponding to the crossover to the tunnelling regime

conductance are roughly located near integer multiples of G_0. This step-wise structure is a clear indication of the formation of point contacts in the metal electrode before the complete breaking of the formed nanowires. This is similar to what it is seen in experiments done with the other techniques explained in Sect. 1.6.1.3 [7]. The long experimental time to create one of these constrictions (hundreds of seconds) compared to other techniques such as STM or MBJ (tens of microseconds) does not favor the building of histograms with enough statistics to assign the observed step-wise behavior to conductance quantization. From tens of experiments, we can conclude the reproducible existence of steps in the conductance on the verge of the metal-tunneling crossover and certain dispersion in the specific values of the conductance plateaus, which we always observe below 4 G_0. In any case, previous studies indicate that conductance histograms are more irregular in transition metal nanoconstrictions [7, 8] than in the case of monovalent metals such as Au [9]. This is due to the fact that the number of conducting channels in a single-atom contact is determined by the number of valence electrons and their orbital state. Theoretical works show that the conductance in non-monovalent metals strongly depends on the specific atomic configurations of the constriction [10–12], which suggests that strong dispersion in the specific values of the conductance plateaus are expected in a reduced number of experiments. To determine the values where steps in conductance occur, numerical differentiation of the G(t) curve was done. Peaks with approximately zero maximum value are observed where plateaus in G occur.

In the particular case of Fig. 4.3, the fit of these peaks to a Gaussian curve gives the maximum at the values $(1.02 \pm 0.01)\ G_0$, $(1.72 \pm 0.01)\ G_0$, $(2.17 \pm 0.01)\ G_0$, and $(3.24 \pm 0.02)\ G_0$. The broadness of the identified steps is a typical feature of room temperature measurements [7, 8, 13].

In Fig. 4.4, SEM images taken during the etching process are shown. The etching by Ga ions in the first stages of milling is highly inhomogeneous due to the presence of an Al layer deposited deliberately on top of the Cr layer. The granular

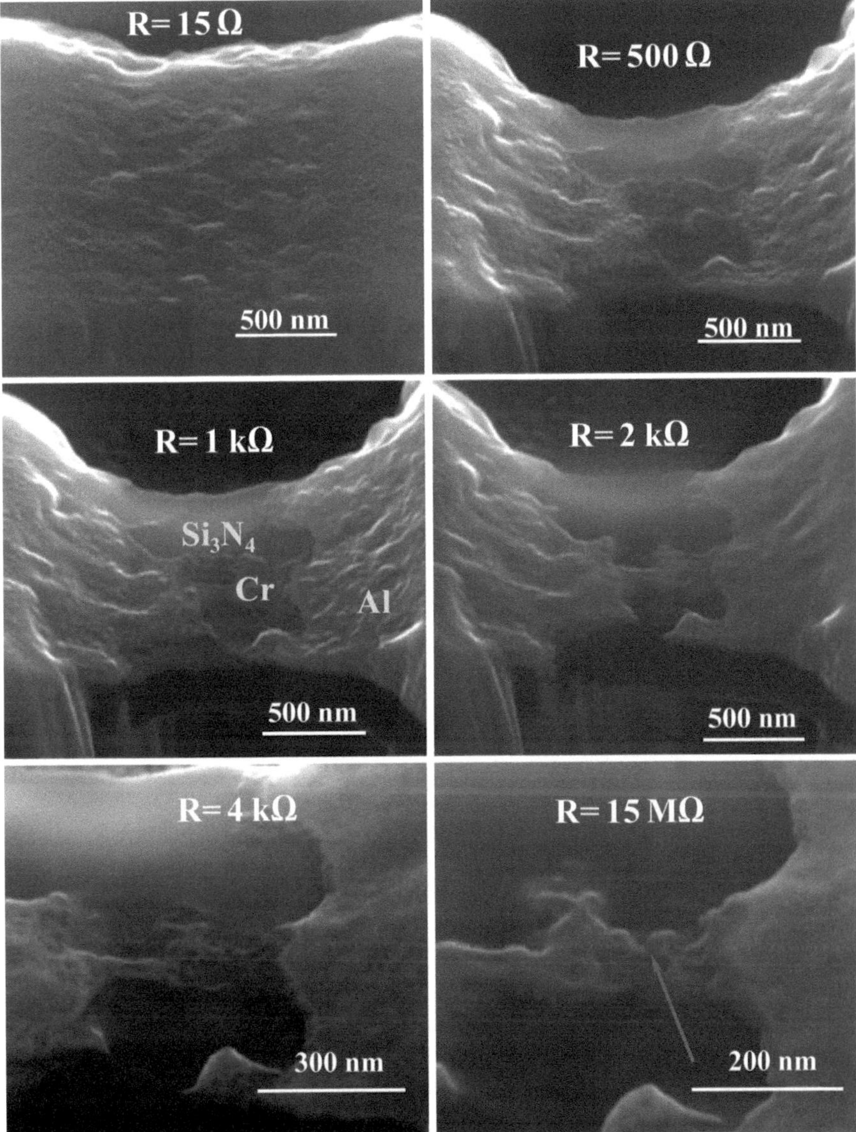

Fig. 4.4 SEM images at different stages of milling (view at 52° tilt angle), showing the change of the microstructure of the Al/Cr electrode as a function of time. The high magnification used in the images allows the visualization of the evolution of the material under ion bombardment at a sub-micrometric scale. The resistance value in each case is shown. On the verge of the metal-tunnel crossover (image corresponding to R = 4 kΩ) the current flows through a single nanoconstriction and atomic-sized-conduction features are observed. When the last nanoconstriction becomes fractured (image corresponding to R = 15 MΩ), the resistance jumps to MΩ range values. The arrow signals the breaking point of the constriction

morphology presented by the evaporated aluminum favors this phenomenon, known as "channelling", in which preferential sputtering of some grains with respect to others with different crystalline orientation occurs [14]. At this milling-time, numerous micrometric channels conduct the current. The progressive thinning during the process as well as the fracture of some of them involves a progressive and continuous increase in resistance. At around 1 kΩ one can clearly distinguish how the Al layer has been fully removed from the constriction zone. The chromium is now solely responsible for conduction. The Cr grain size is smaller than in Al, and a more homogeneous etching process takes place from that moment on. The plateaus found for the conductance correspond to images where only one metal nanowire is not fractured, reaching atomic-sized contacts just before the final breaking and the crossover to the tunneling regime. When this finally occurs, an abrupt jump occurs in the resistance, reaching values in the MΩ range, associated with a tunneling conduction mechanism.

In some constrictions, the etching was stopped at different moments, measuring I–V curves. As can be seen in Fig. 4.5, the linear I–V behavior indicates

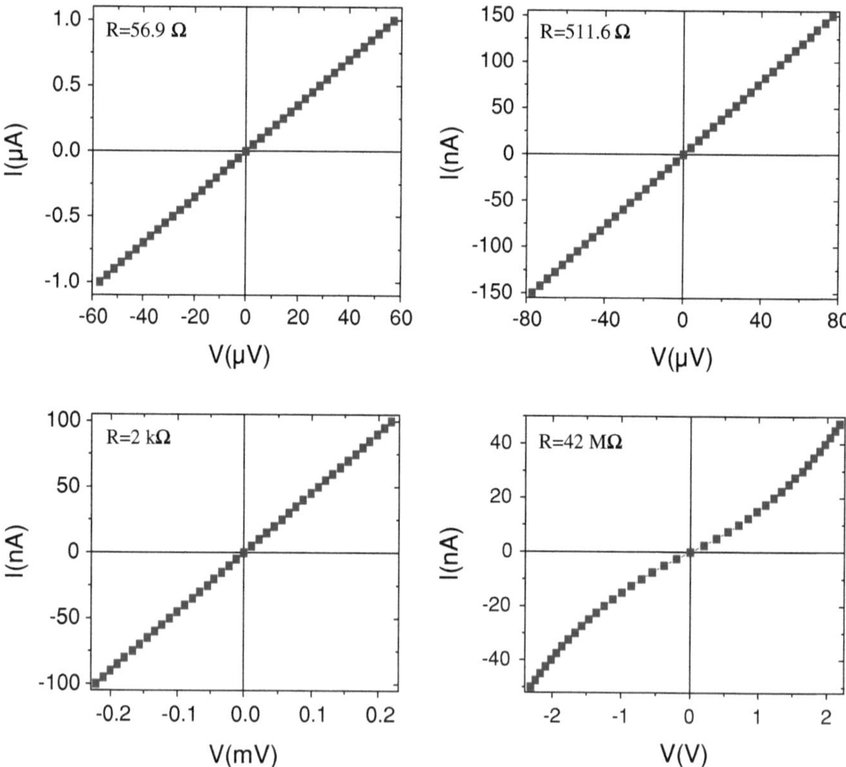

Fig. 4.5 I–V curves measured at fixed resistance values. A linear dependence, as expected for metallic conduction is observed except for the last stages of milling, where the tunnelling regime has been reached

metallic conduction, and consequently that the constriction is not broken. Once the constriction breaks, a non-linear I–V behavior is found, as expected in tunneling conduction. Fitting the I–V curves for tunneling conduction to the Simmons model for a rectangular tunnel barrier [15] gives as a result a barrier height of the order of 4.5 V, a junction effective area of 300 nm^2 and a barrier width of the order of 0.65 nm. This barrier height corresponds with the work function for chromium [16] and the other parameters seem reasonable, evidencing the adequacy of the fit in spite of the simplified assumptions of the model. Even though re-sputtering of some Al atoms towards the contact area could be possible, the value found for the barrier height in the tunneling regime seems to indicate that the electronic transport through the nanocontact is governed by the Cr atoms.

We have not observed degradation during the measurement time (tens of seconds) since the substrate under the formed metal constriction guarantees a robust and stable structure. We have found that when the constriction shows resistance values of the order of the conductance quantum, a limiting maximum power of 0.3 nW is required to avoid degradation, which constrains the maximum applied voltage to values below $\sim$6 mV. Considering a contact area of 0.3 $\times$ 0.3 nm^2, a 5.5 $\times$ 10^7 A/cm^2 current density is obtained, which is around three orders of magnitude larger than the maximum current bearable by a macroscopic metallic wire. This huge current density is understood in terms of ballistic transport through the contact [7].

With the available data, it is not possible to discard the possibility, due to the moderate vacuum level (10^{-6} torr) and the Ga-assisted etching process, some impurity could be present in the surroundings of the nanoconstriction and could exert some influence on the conduction properties. However, the linearity obtained in the I–V curves in the metallic regime seems to confirm the presence of clean contacts, because only the presence of adsorbates and contaminants produce non-linear curves in the low-voltage regime [17, 18]. In our case, the use of low ion energy (5 kV) and current (2 pA) is expected to minimize the role of Ga damage and implantation. We have carried out a chemical analysis of the surroundings of the created nanoconstrictions by means of EDX measurements. The results give a typical composition for Si/O/Cr/Ga/C (in atomic percentage) of 30/45/23/1/1. If the substrate elements (Si and O in this case) are discarded, the Cr and Ga atomic percentages are found to be around 96 and 4%, respectively. However, one should be aware that this technique has a very limited resolution in depth and lateral size.

We would like to remark that ion-etching processes are commonly used in micro/nano-fabrication steps in laboratory and industrial processes without affecting the device properties. Thus, it is reasonable to assume that for many types of devices the use of suitable ion energy and dose will not affect the device properties.

Finally, we must point out the importance of the chosen parameters, crucial for observing the transition from the ohmic to the tunneling regime in a detailed way (this is not the case under other conditions) [19].

4.3 Iron Nanocontacts

4.3.1 Creation of Fe Nanoconstrictions Inside the Chamber

The procedure used in this case is approximately the same as that explained in Sect. 4.2 for chromium, but in this case a 10 nm-thick iron layer was deposited before optical lithography and etching, once it was checked that the etching process was inhomogeneous enough to create appropriate structures which favor the formation of constrictions.

In Fig. 4.6, we show a typical example performed with iron. During the first 2 min, the Ga etching process is performed using 5 kV and 70 pA, the resistance approaching 2 kΩ at the end. We then continue the Ga etching process using 5 kV and 0.15 pA. The process is slower and allows the detection of step-wise changes in the resistance before the jump to the tunneling regime, similar to the observed behavior in the case of the Cr-based electrodes. The inset shows a close-up of the same results representing the conductance (in units of G_0) versus time. Clear steps are observed which are also interpreted as evidence of the formation of atomic-size metallic contacts just before entering the tunneling regime. These same features were observed for tens of samples, indicating that the process is suitable for iron. The constrictions are found to be stable inside the vacuum chamber. The next step was to measure MR effects of the created constrictions, involving the opening of the vacuum chamber and putting the sample in contact with the atmosphere.

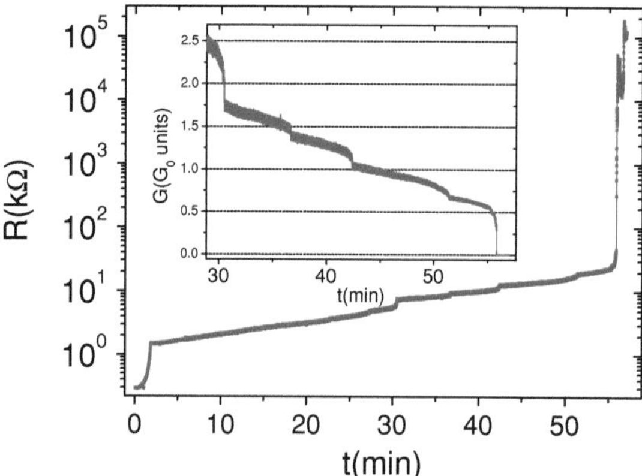

Fig. 4.6 Resistance versus time in a typical ion etching process of an Fe electrode. The *inset* shows the corresponding conductance (in G_0 units) versus etching time. In this stage of milling, steps of the order of some fraction of G_0 are seen, corresponding to discrete thinning of the contact area of the order of one atom. The last step is followed by a sharp decrease in conductance, corresponding to the crossover to the tunneling regime

4.3.2 Measurement of One Constriction in the Tunneling Regime of Conduction

As an example of the typical problems found when Fe contacts are exposed to ambient conditions, we show in this section the MR results for one sample.

4.3.2.1 Creation of the Constriction

In this particular case, the Ga-FIB etching was done in a region of around $2 \times 6 \ \mu m^2$ (square in Fig. 4.7a), with a beam acceleration voltage of 5 kV and a beam current of 10 pA, constant through the process. The etching process took a few minutes time, roughly corresponding to a total ion dose of 10^{17} ions/cm^2. In Fig. 4.7b, the microstructure of the constriction after the milling is shown.

In Fig. 4.8, we show the evolution of the resistance with the process time. Since the initial resistance of the iron electrode is $R_i \sim 3 \ k\Omega$, the resistance corresponding to the etched part (R_c) is the measured value minus R_i ($R = R_i + R_c$). After 2 min of etching, R starts to increase abruptly, and the FIB column is stopped when $R_c \approx 8 \ k\Omega$. The R is measured during several minutes, finding that the constriction, in the metallic regime, is stable under the high vacuum conditions of the chamber ($P \sim 10^{-6}$ mbar). To avoid the possible deterioration when exposed to ambient conditions, a ~ 10 nm-thick layer of Pt–C was deposited by electrons (FEBID) on top of the etched zone (one single deposition, at $0°$), using $(CH_3)_3Pt(CpCH_3)$ as gas precursor. The high resistance of this material due to the high amount of carbon guarantees a resistance in parallel to the constriction of the order of tens of $M\Omega$ (see Sect. 5.2 for details), which is confirmed by the negligible change of R while the deposit is done.

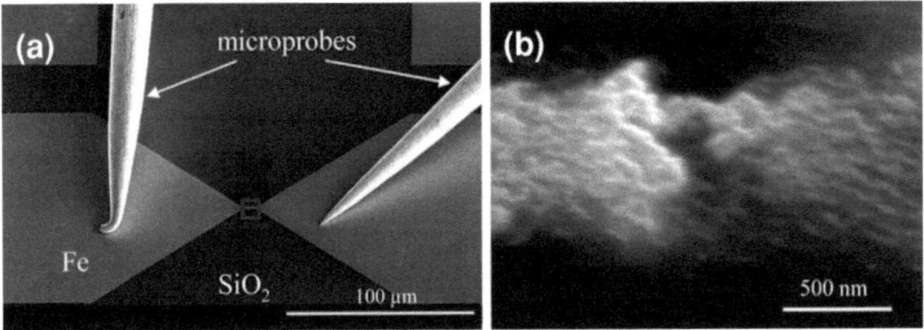

Fig. 4.7 $52°$ tilted-view SEM images of the fabricated constriction. **a** Experimental configuration, with the two microprobes contacted to the Fe pads for the in situ control of the resistance while the process is taking place. The *square* indicates the etched zone. **b** Microstructure of the iron after etching. A constriction is formed as a consequence of the ionic bombardment

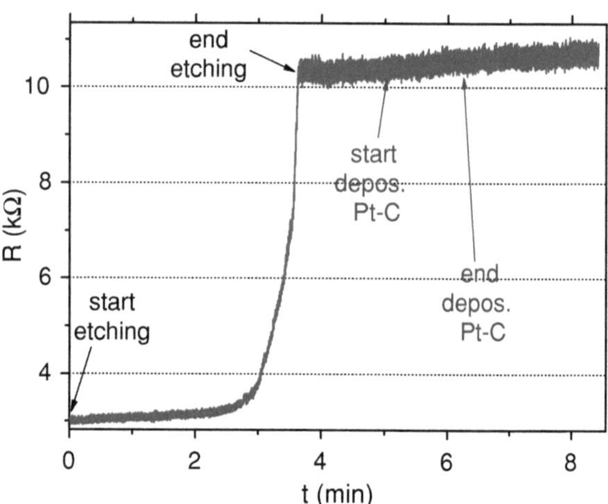

Fig. 4.8 Resistance of the iron electrode as a function of the process time. The FIB etching is stopped when the resistance reaches 11 kΩ ($R_c \sim 8$ kΩ). The constriction is covered by a thin layer of Pt–C, deposited by FEBID. The fabricated nanostructure is stable inside the vacuum chamber

4.3.2.2 Magnetoresistance Measurements

Once the constriction had been fabricated, the sample was exposed to ambient conditions over a few minutes, and transferred to the CCR-electromagnet system explained in Chap. 2. The measurements of the resistance at room temperature showed an increase in the resistance by a factor of 10 ($R \approx 100$ kΩ), evidencing the departure from metallic conduction in the nanoconstriction. This value is significantly above the resistance corresponding to the quantum of conductance ($R_0 = 1/G_0 = 12.9$ kΩ). The non-linearity of the current-versus-voltage curves (not shown here) indicates that the constriction is no longer in the metallic but in the tunneling regime. The high reactivity of the nanostructure due to its large surface-to-volume ratio seems to be the most reasonable explanation of this effect. The resistance increased up to 120 kΩ at $T = 24$ K.

We measured the magnetoresistance at low temperatures with the magnetic field parallel to the current path (LG). In Fig. 4.9, the evolution with H at several temperatures is shown. MR ratios of the order of 3.2% are obtained. This value is a factor 30 times higher than in Fe non-etched samples, used as reference, where we observed a MR = −0.11% in this configuration (note the different sign). As is schematically explained by the arrows in the graph, the evolution of R is understood by a change from a parallel (P) to an anti-parallel (AP) configuration of the ferromagnetic electrodes, separated by an insulator. In greater detail, starting from saturation, a first continuous increase in R is observed, of the order of 0.8%. This seems to be caused by a progressive rotation of the magnetization (M) in one of the electrodes. At around $H = 700$ Oe, an abrupt jump of the resistance occurs (MR = 2.5%), since the magnetization of one electrode switches its direction, and almost aligns AP to the M of the other electrode (intermediate state: IS). The MR becomes maximum at $H = 1$ kOe, when the magnetization in both electrodes is AP. At $H \approx 2$ kOe, the M in the hard electrode also rotates, resulting again in a low-R state (P). As T increases, the IS, previously explained, disappears. IS is

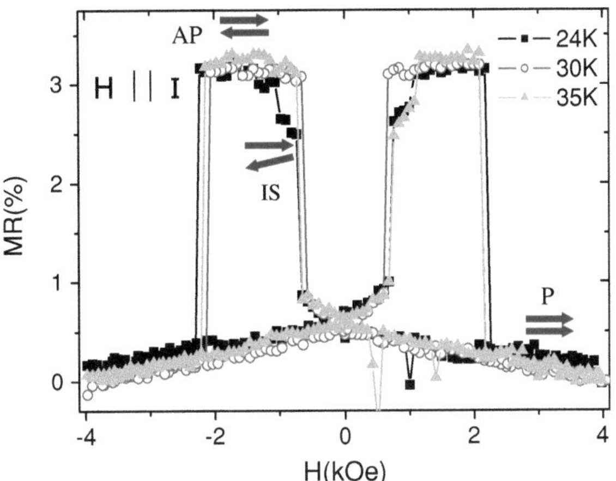

Fig. 4.9 MR as a function of the magnetic field at low temperatures. The MR is positive, with a value of 3.2% at 24 K, and can be understood by the decoupling of the magnetic electrodes separated by a tunneling barrier

probably caused by the pinning of the M by some defects present in the soft electrode. Thus, the increase in the thermal energy favors the depinning of M. When the temperature was increased above 35 K, the nanoconstriction became degraded, probably due to the flow of the IS electrical current.

In Fig. 4.10, the dependence of the MR at 24 K as a function of the applied voltage is shown, for the P and AP configurations. The diminishment of the resistance with the voltage is a typical feature of MTJs, attributed to several factors such as the increase of the conductance with bias, excitation of magnons, or energy dependence of spin polarization due to band structure effects [20]. As the magnetostrictive state of parallel and antiparallel electrodes is the same, it seems that magnetostriction is not the cause of the observed TMR effect.

We have also studied the dependence of the anisotropy of the MR with the field angle at saturation, by rotating the sample at the maximum field attained, $H = 11$ kOe. In Fig. 4.11, the evolution of MR is shown as a function of θ, the

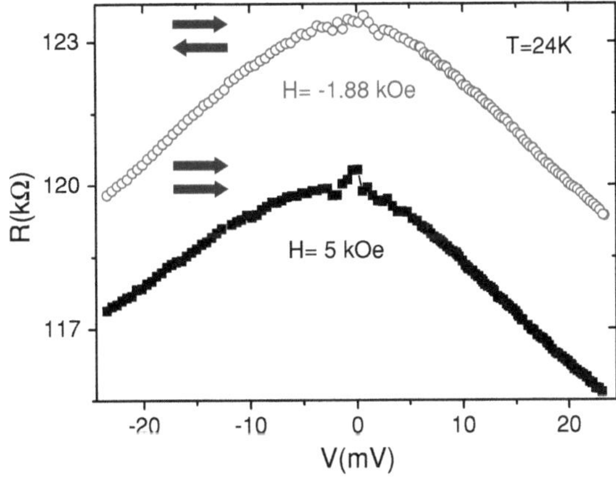

Fig. 4.10 Bias dependence of the magnetoresistance at $T = 24$ K for AP ($H = -1.88$ kOe) and P ($H = 5$ kOe) configurations

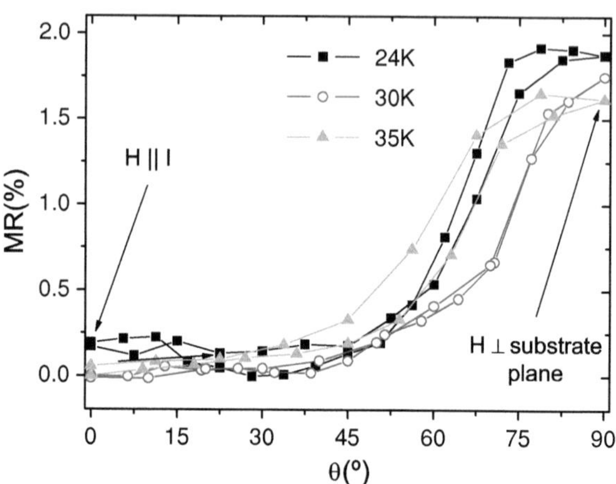

Fig. 4.11 Angle dependence of the MR for low temperatures. A fixed field, H = 11 kOe, is applied

angle formed between H and the substrate plane. An AMR effect is present in the tunneling regime (TAMR). This effect is around 2% at 24 K, higher and of different sign from the AMR occurring in the bulk material ($\sim -0.3\%$). As remarked in Sect. 1.6.2, a TAMR has been previously observed in iron nanocontacts fabricated by MBJ [21], and implies that the evanescent wave functions maintain a strong atomic orbital character. The angle dependence is found to be more abrupt than the normal one for the AMR, proportional to $\cos^2\theta$. This behavior is typical for BAMR [21, 22], and can be understood by considering the details of orbital overlap [23, 24].

The experiments as a function of angle are also helpful to discard the possibility that magnetostriction effects are responsible for the observed MR. The magnetostriction in iron is given by:

$$\frac{\Delta l}{l} = \frac{3}{2}\lambda_s\left(\cos^2\theta - \frac{1}{3}\right) \tag{4.1}$$

with $\Delta l/l = -7 \times 10^{-6}$ for $\theta = 0°$, and $\Delta l/l = 3.5 \times 10^{-6}$ for $\theta = 90°$.

Thus, the increase of the MR as a function of θ has the opposite sign than if it was caused by magnetostriction, where the minimum resistance would be expected when H is perpendicular to the substrate plane.

The results for this sample provide an example of the behavior of all the samples we have created. Although it is possible to fabricate Fe nanoconstrictions near the ballistic regime of conductance by the method developed, which are stable inside the FIB chamber, they are degraded when exposed to ambient conditions, breaking or oxidizing, and changing to the tunneling regime of conductivities. For the sample shown here, the TMR ratios at low temperature are 30 times higher than non-etched samples. We also observe a TAMR effect at low temperatures, of the order of 2%, and with an angle dependence different from the $\cos^2\theta$ expected for bulk AMR. From these experiments we ensure that magnetostriction does not play any role in the measurements.

4.4 Conclusions

A new approach to obtain stable atomic-sized contacts at room temperature under high vacuum has been performed. The combination of FIB etching with simultaneous electrical measurements allows a fine control of the resistance. The steps in the conductance immediately prior to the crossover to the tunnelling regime are an indication of atomic reconstructions. The plateaus at different conductance values are associated with different atomic configurations in the constriction. The simultaneous imaging of the etching process via the electron column allows the structure formed to be followed in real-time.

The constrictions are stable at room temperature and under high vacuum conditions, proved for Cr and Fe. The possibility of establishing point contacts in a wide variety of materials under similar conditions is expected with the methodology described. This is not the case for other techniques, which are less flexible with respect to the materials and protocols used. In our case, the sputtering rate of the material under etching at the ion energy used seems to be the most relevant factor together with the ion column current, which influences the ion beam width and the etching rate. We can foresee a great potentiality in this method for the creation of functional devices in the field of Spintronics.

In the case of Fe nanocontacts, the difficulty to have stable nanocontacts when exposed to ambient conditions has been shown. In spite of the protective covering, the constrictions are either completely degraded, or switch to the tunneling regime when coming in contact with air.

We have shown an example of one iron contact showing tunneling conduction. This has TMR ratios at low temperature 30 times higher than non-etched samples. We also observe a TAMR effect at low temperatures, of the order of 2%, and with an angle dependence different from the $\cos^2\theta$ expected for bulk AMR. From these experiments we also ensure that magnetostriction does not play any role in the measurements. This result shows that under low voltage etching, and with a moderate ion dose, the FIB procedure does not destroy the magnetic properties of the devices, although more research is required to investigate to what extent they are affected. The high stability expected for these constrictions in comparison with other suspended atomic-size structures makes this result promising for future research. A systematic study of MR is currently in progress in order to improve the stabilization of constrictions in the metallic range of resistances, and investigate if high BMR and BAMR values can be attained at room temperature, which would have a high impact in the field.

We should also remark that a different approach has been developed in our laboratory during this thesis, following the same philosophy as the method described in this chapter. The main advantage is the well-defined current direction attained, suitable for anisotropic magnetoresistance measurements. However, it is performed at 30 kV, inducing a higher amount of amorphization [14], with a possible increase in the degree of deterioration of the magnetic properties. For details, see Oboňa et al. [25].

References

1. N. García, C. Hao, L. Yonghua, M. Muñoz, Y. Chen, Z. Cui, Z. Lu, Y. Zhou, G. Pan, A.A. Pasa, Magnetoresistance in thin permalloy film (10 nm thick and 30–200 nm wide) nanocontacts fabricated by e-beam lithography. Appl. Phys. Lett. **89**, 083112 (2006)
2. P. Krzysteczko, G. Dumpich, Magnetoresistance of Co nanoconstrictions fabricated by means of electron beam lithography. Phys. Rev. B **77**, 144422 (2008)
3. T. Huang, K. Perzlaimer, C.H. Back, In situ magnetoresistance measurements of ferromagnetic nanocontacts in the Lorentz transmission electron microscope. Phys. Rev. B **79**, 024414 (2009)
4. O. Céspedes, S.M. Watts, J.M. Coey, K. Dörr, M. Ziese, Magnetoresistance and electrical hysteresis in stable half-metallic $La_{0.7}Sr_{0.3}$ MnO_3 and Fe_3O_4 nanoconstrictions. Appl. Phys. Lett. **87**, 083102 (2005)
5. S. Khizroev, Y. Hijazi, R. Chomko, S. Mukherjee, R. Chantell, X. Wu, R. Carley, D. Litvinov, Focused-ion-beam-fabricated nanoscale magnetoresistive ballistic sensors. Appl. Phys. Lett. **86**, 042502 (2005)
6. I.V. Roshchin, J. Yu, A.D. Kent, G.W. Stupian, M.S. Leung, Magnetic properties of Fe microstructures with focused ion beam-fabricated nano-constrictions. IEEE Trans. Magn. Mat. **37**, 2101 (2001)
7. N. Agraït, A.L. Yeyati, J.M. Van Ruitenbeek, Quantum properties of atomic-sized conductors. Phys. Rep. **377**, 81 (2003)
8. C. Sirvent, J.G. Rodrigo, S. Vieira, N. JurczyszynMingo, F. Flores, Conductance step for a single-atom contact in the scanning tunneling microscope: noble and transition metals. Phys. Rev. B **53**, 16086 (1996)
9. E. Scheer, N. Agraït, J.C. Cuevas, A.L. Yeyati, B. Ludoph, A. Martín-Rodero, G.R. Bollinger, J.M. van Ruitenbeek, C. Urbina, The signature of chemical valence in the electrical conduction through a single-atom contact. Nature **394**, 154 (1998)
10. J.J. Palacios, A.J. Pérez-Jiménez, E. Louis, E. SanFabián, J.A. Vergués, First-principles approach to electrical transport in atomic-scale nanostructures. Phys. Rev. B **66**, 035322 (2002)
11. A. Hasmy, A.J. Pérez-Jiménez, J.J. Palacios, P. García-Mochales, J.L. Costa-Kramers, M. Díaz, E. Medina, P.A. Serena, Ballistic resistivity in aluminum nanocontacts. Phys. Rev. B **72**, 245405 (2005)
12. F. Pauly, M. Dreher, J.K. Viljas, M. Häfner, J.C. Cuevas, P. Nielaba, Theoretical analysis of the conductance histograms and structural properties of Ag, Pt, and Ni nanocontacts. Phys. Rev. B **74**, 235106 (2006)
13. D.R. Strachan, D.E. Johnston, B.S. Guitom, S.S. Datta, P.K. Davies, D.A. Bonell, A.T.C. Johnson, Real-time TEM imaging of the formation of crystalline nanoscale gaps. Phys. Rev. Lett. **100**, 056805 (2008)
14. L. Giannuzzi, F. Stevie, *Introduction to Focused Ion Beams: Instrumentation, Techniques, Theory and Practices* (Springer Science Business Media, Boston, 2005)
15. J.G. Simmons, Generalized formula for the electric tunnel effect between similar electrodes separated by a thin insulating film. J. Appl. Phys. **34**, 1793 (1963)
16. H.B. Michaelson, The work function of the elements and its periodicity. J. Appl. Phys. **48**, 4729 (1977)
17. K. Hansen, S.K. Nielsen, M. Brandbyge, E. Laegsgaard, I. Stensgaard, F. Besenbacher, Current-voltage curves of gold quantum point contacts revisited. Appl. Phys. Lett. **77**, 708 (2000)
18. K. Hansen, S.K. Nielsen, E. Laegsgaard, I. Stensgaard, F. Besenbacher, Fast and accurate current-voltage curves of metallic quantum point contacts. Rev. Sci. Instrum. **71**, 1793 (2000)
19. A. Fernández-Pacheco et al., Nanotechnology **19**, 415302 (2008) (for more processes and http://iopscience.iop.org/0957-4484/19/41/415302/media for a video of the formation of a constriction)

20. J.S. Moodera, G. Mathon, Spin polarized tunneling in ferromagnetic junctions. J. Magn. Mag. Mat. **200**, 248 (1999)
21. M. Viret, M. Gauberac, F. Ott, C. Fermon, C. Barreteau, G. Autes, R. Guirado López, Giant anisotropic magneto-resistance in ferromagnetic atomic contacts. Eur. Phys. J. B **51**, 1 (2006)
22. A. Sokolov, C. Zhang, E.Y. Tsymbal, J. Redepenning, B. Doudin, Quantized magneto-resistance in atomic-size contacts. Nat. Nanotechnol. **2**, 171 (2007)
23. J. Velev, R.F. Sabirianov, S.S. Jaswal, E.Y. Tsymbal, Ballistic anisotropic magneto-resistance. Phys. Rev. Lett. **94**, 127203 (2005)
24. D. Jacob, J. Fernández-Rossier, J.J. Palacios, Anisotropic magnetoresistance in nanocontacts. Phys. Rev. B **77**, 165412 (2008)
25. V. Oboňa, J.M. de Teresa, R. Córdoba, A. Fernández-Pacheco, M.R. Ibarra, Creation of stable nanoconstrictions in metallic thin films via progressive narrowing by focused-ion-beam technique and in situ control of resistance. Micr. Eng. **86**, 639 (2009)

Chapter 5
Pt–C Nanowires Created by FIBID and FEBID

In this chapter we will present a thorough characterization of the electrical properties of Pt–C nanowires created by FIBID and FEBID. The same methodology as in Chap. 4 will be used, where the resistance of the NWs is probed as the NW is being grown, getting a perfect control of the process. The electrical properties are studied at room temperature, "in situ", and as a function of temperature, "ex situ". These results are correlated with chemical and microstructural characterization, and can be understood within the framework of the theory by Mott and Anderson for disordered materials.

The chapter is mainly devoted to the study of wires created by FIB. By depositing NWs with different thicknesses, we study deposits in different ranges of conductivity. These NWs present a metal–insulator transition as a function of thickness, with the metal–carbon proportion as the key parameter that determines the mechanism for electrical conduction.

Besides, this same methodology has been used to compare NWs grown by FIBID in the metallic regime of conduction with wires grown by FEBID. The deposits induced by electrons are orders of magnitude more resistive than those induced by ions for all thicknesses, resembling Pt–C FIBID NWs in the insulating regime of conduction. The substantial differences in the electrical transport measurements between both types of deposits can be understood by means of the theories detailed for FIBID-NWs, giving rise to a complete picture for Pt–C deposits grown by focused beams.

5.1 Nanowires Created by Focused-Ion-Beam-Induced-Deposition

5.1.1 Previous Results in Pt–C Nanodeposits Grown by FIBID

One of the most commonly deposited metallic materials using FIBID is Pt. A lot of work has been done regarding its electrical transport properties, with a wide range

A. Fernandez-Pacheco, *Studies of Nanoconstrictions, Nanowires and Fe₃O₄ Thin Films*, Springer Theses, DOI: 10.1007/978-3-642-15801-8_5,
© Springer-Verlag Berlin Heidelberg 2011

of reported results on the resistivity at room temperature (ρ_{RT}), as well as on the thermal dependence of the resistance. We show in Table 5.1. the results compiled in that work, together with additional references [1–12]. In addition to the value for ρ_{RT}, we also include, when reported, the thermal coefficient of the resistivity ($\beta = d\rho/dT$), the residual resistivity ratio (RRR $= \rho_{RT}/\rho_{lowT}$), the electronic mechanism proposed for conduction, the character of current-versus-voltage curves, the chemical composition of the deposits, and the gas precursor used. As revealed by TEM, the microstructure found rather inhomogeneous, with crystalline metallic inclusions (around 3–10 nm) embedded in an amorphous carbon matrix [2–6]. An important point that is usually missed when comparing these results is the two types of precursors used for Pt deposition. We can see that in two groups [3–5 and 10], the nanowires (NWs) are deposited with cyclopentadienyl-trimethyl platinum, $(CH_3)_3PtCp$, whereas in the rest of groups [1, 2, 6–9, 11, 12] methyl-cyclopentadienyl-trimethyl platinum, $(CH_3)_3Pt(CpCH_3)$ is used. Deposits using the first precursor (one carbon less present in the molecule) have the lowest resistivity reported (only about 7 times higher than that of bulk Pt, ~ 10.8 $\mu\Omega$ cm) and a positive β. At low temperatures they present deviations from the behavior expected for pure Pt NWs, such as weak antilocalization and quasi-one- dimensional-interference effects [3–5]. On the contrary, with the second precursor, β is always negative, either with an appreciable thermal dependence, associated with variable-range hopping (VRH) [1, 12], or an almost-negligible dependence [2, 9], resembling the conduction of a dirty metal, with metallic [6] or tunneling conduction [2]. In this work, we center our study on NWs created with the second precursor, of more extended use in the scientific community, with the aim to clear up the, in principle, contradictory reported results.

5.1.2 Experimental Details

5.1.2.1 Deposition Parameters

The Pt–C NWs were grown using the FIB column integrated in the Dual Beam system, together with the GIS to introduce the gas into the process chamber, all explained in detail in Sect. 2. The SEM was used for imaging the sample, minimizing the ion dose in the deposits. The deposition parameters are: $(CH_3)_3Pt(CpCH_3)$ precursor gas, $(T_G) = 35°C$ precursor temperature, $(V_{FIB}) = 30$ kV beam voltage, $(I_{FIB}) = 10$ pA beam current, substrate temperature $= 22°C$, dwell time $= 200$ ns, chamber base pressure $= 10^{-7}$ mbar, process pressure $= 3 \times 10^{-6}$ mbar, beam overlap $= 0\%$, distance between GIS needle and substrate $(L_D) = 1.5$ mm. Under these conditions, a dose of 3×10^{16} ion/cm^2 min roughly irradiates the sample. We must emphasize that the values for T_G (about 10°C below the usual operation temperature) and L_D (~ 1.35 mm higher than usually) were chosen to decrease the deposition rate, allowing detailed measurements of the evolution of the resistance as the deposit was realized. This is required to get fine

Table 5.1 Compilation of reports for Pt–C NWs created by FIB

Report	ρ_{RT} ($\mu\Omega$ cm)	$\beta=$ dρ/dT	RRR= ρ_{RT}/ρ_{lowT}	Conduction	I–V curves	Atomic % C:Pt:Ga:O	Gas precursor
Lin et al. [3–5]	61.5–545	>0	≈1.3	Disordered metal	Linear	67:30:3:0	$(CH_3)_3PtCp$
Tao et al. [6]	70–700	–	–	–	–	24:46:28:2	$(CH_3)_3Pt(CpCH_3)$
Puretz et al. [7]	400–2100	–	–	–	–	47:37:13:0	$(CH_3)_3Pt(CpCH_3)$
Telari et al. [8]	500–5000	–	–	–	–	70:20:10:0	$(CH_3)_3Pt(CpCH_3)$
De Teresa et al. [9]	800	<0	0.8	Disordered metal	Linear	70:20:10:0	$(CH_3)_3Pt(CpCH_3)$
Peñate-Quesada et al. [2]	860–3078	<0	≈1	Inter-grain Tunneling	Non-linear	Similar to 210–212	$(CH_3)_3Pt(CpCH_3)$
Tsukatani et al. [10]	1000	≈0	≈1	–	–	–	$(CH_3)_3PtCp$
Dovidenko et al. [11]	1000–5000	–	–	–	–	63:31:4:2	$(CH_3)_3Pt(CpCH_3)$
De Marzi et al. [12]	2200	<0	≈0.7	VRH	–	–	$(CH_3)_3Pt(CpCH_3)$
Langford et al. [1]	1000–10^6	<0	0.6	VRH	Linear	50:45:5:5	$(CH_3)_3Pt(CpCH_3)$
Present study	700–10^8	<0	Thickness dependent	MIT	Thickness dependent	Thickness dependent	$(CH_3)_3Pt(CpCH_3)$

All deposits are done at 30 kV, except in 5.6 (30–40 kV) and 5.7 (25 kV). In these two references the atomic percentage is obtained by Auger spectroscopy, whereas the rest are EDX measurements. Contradictory results are fund for the transport properties and composition. The present work can explain the differences in terms of a different composition with thickness. In ref 5.2, the non-linear I–V characteristics were observed for a 3078 $\mu\Omega$ cm NW

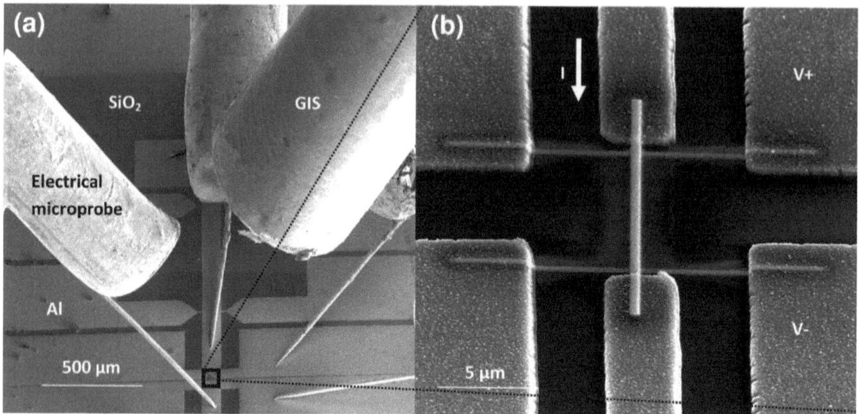

Fig. 5.1 **a** SEM image of the experimental configuration for 4-wire electrical measurements. The 4 microprobes are contacted to Al micrometric pads, while the deposit is carried out. For deposition, the GIS needle is inserted near the substrate. **b** SEM image of one deposited Pt–C NW (*vertical line*). The additional two horizontal lines are done before the vertical NW, to perform four wire measurements (see text for details)

control, especially during the initial stages, where a very abrupt change in the resistance occurs in a very short period of time (see Sect. 5.3). We obtained approximately the same final resistivity when depositing NWs using normal conditions (see Sect. 5.2.1.), so these changes do not seem to affect substantially the properties of the NWs. Thermally-oxidized silicon (~ 200 nm of SiO_2) was used as substrate, and aluminum pads were previously patterned by optical lithography (see lift-off process in Sect. 2.1.1 for details). The studied NWs (top–bottom deposit in Fig. 5.1b) are length (L) = 15 µm, width (w) = 500 nm, thickness (t) = variable with process time.

5.1.2.2 "In Situ" Measurement of the Resistance During the Growth Process

The electrical resistance was measured inside the process chamber using the experimental setup explained in detail in Sect. 2.2.2. In Fig. 5.1a, a SEM image shows the experimental configuration for the deposition, where the 4 microprobes are contacted to the pads, and the GIS needle is inserted near the substrate. By applying a constant current between the two extremes of the NW, and measuring the voltage drop in two of the intermediate pads, we are able to measure the resistance of the NW which is being grown, once it is below 1 GΩ. This method is similar to others reported in literature [13, 14]. The probe current (I_{probe}) was changed during the measurement, in order to optimize the signal-to-noise ratio, and trying to minimize heating effects of the device while it was created. Typical values for I_{probe} range from 5 nA at the beginning of the monitoring (R $\sim$ 1 GΩ)

up to 10 μA at the end (R $\sim$ 1 kΩ). All these measurements have been done at room temperature

In Fig. 5.1b the vertical deposit is an example of a studied NW. Before its deposition, additional "perpendicular-to-the-NW" deposits were performed (horizontal deposits) for four wire measurements, avoiding non-linear effects associated with the resistance of the contacts [15, 16]. These extra-connections were done using several growth parameters, finding no differences in the resistance measurements of the NWs. This avoids any influence of the associated halo to the deposition in the results, which could be the source of spurious effects [11, 17].

5.1.2.3   Compositional Analysis by EDX

For the study of the composition by Energy Dispersive X-Ray (EDX), we performed deposits on SiO_2 of the same lateral dimensions as those used for electrical measurements, and varying thickness. The EDX measurements were done by means of a commercial Oxford INCA 200 EDX setup whose detector is driven in the vicinity of the sample. The selected energy for the microanalysis is 20 kV. Prior to each EDX experiment, energy calibration by means of a Co calibration sample was done.

5.1.2.4   Structural Analysis via Scanning-Transmission-Electron-Microscopy

Cross-section lamella of selected NWs was performed with the dual beam equipment by means of Ga ion thinning initially at 30 kV and finally at 5 kV. An Omniprobe nanomanipulator was used to perform in situ lift-out and placing of the lamella on a Cu grid. The STEM images were obtained inside the same experimental setup, in Dark Field mode, under 30 kV and 0.15 nA conditions.

5.1.2.5   XPS Measurements

In order to study the chemical nature of the deposits, we have done depth-profile XPS measurements the equipment explained in Sect. 2.3.2, at a base pressure of 3×10^{-10} Torr. Because of the limited spatial resolution of the X-ray probe, this deposit was much larger in comparison with all the others: length = 100 μm × width = 100 μm × thickness = 200 nm. Under the chosen conditions, previously detailed, this would imply a process time of about 25 h, so we changed the I_{BEAM} to 3 nA, reducing significantly the deposition time. The dose rate, $\sim 10^{16}$ ions/cm^2-min, was approximately the same in both types of experiments. A 5 kV-argon etching, with current densities of the order of 0.15 A/m^2 was done, to obtain a XPS depth profile of the deposits (probing every 30 nm).

For the quantitative analysis of the XPS spectra, a Shirley-type background was subtracted, using pseudo-Voigt peak profiles with a 10–30% Lorentzian contribution in all edges, except for the Pt 4f. In this case, because of the asymmetry of the peaks, the spectra were analyzed using the Doniach-Sunjic function [18]:

$$I(\varepsilon) = \frac{\Gamma(1-\alpha)}{(\varepsilon^2 + \gamma^2)^{(1-\alpha)/2}} \cos\left(\frac{1}{2}\pi\alpha + \theta(\varepsilon)\right) \tag{5.1.a}$$

$$\theta(\varepsilon) = (1+\alpha)\tan^{-1}(\varepsilon/\gamma) \tag{5.1.b}$$

where ε is the energy measured relative to Fermi energy, Γ is the gamma function, γ is the natural linewidth of the hole state corresponding to its lifetime, and α is the line asymmetry parameter. The function was convoluted with a Gaussian curve with full width at half maximum (FWHM) of 0.3 eV, to take into account the experimental broadening (instrumental resolution, sample inhomogeneity, etc.) [19].

5.1.2.6 Transport Measurements as a Function of Temperature

The measurements of the electrical resistance as a function of temperature were performed in the CCR system detailed in Sect. 2.2.1.

5.1.3 Results

5.1.3.1 Compositional (EDX) and Structural (STEM) Analysis of the Deposits

We performed chemical analysis by EDX on NWs with different thicknesses. The composition was found to be homogeneous over the entire deposited surface, but strongly dependent with deposition time.

The results are shown in Fig. 5.2a. For t = 25 and 50 nm, the deposit is highly rich in carbon: $\sim$68%, with Pt $\sim$ 18% and Ga $\sim$ 14% (atomic percentages). A residual fraction of oxygen (less than 1%) is also present. For higher thickness, the metal content increases gradually, till approximately saturated values: C $\sim$ 40–45%, Pt $\sim$ 30–35%, Ga $\sim$ 25%.

We have fabricated 200 nm-thick lamellae for STEM imaging inside the chamber. We can see an example in Fig. 5.2b, where the difference in dark–bright contrast indicates a gradient in composition as a function of thickness. For the initial deposited nanometers (roughly 50 nm), the image shows a higher C concentration than in the upper part of the deposit, where a higher metal content is present. This agrees perfectly with the former EDX analysis, and was previously found in TEM images by Langford et al. [1].

We can understand the gradient in composition with thickness taking into account two possible factors. First, an important decrease in the secondary electrons yield in SiO_2 that occurs when this oxide is irradiated with ions in the keV

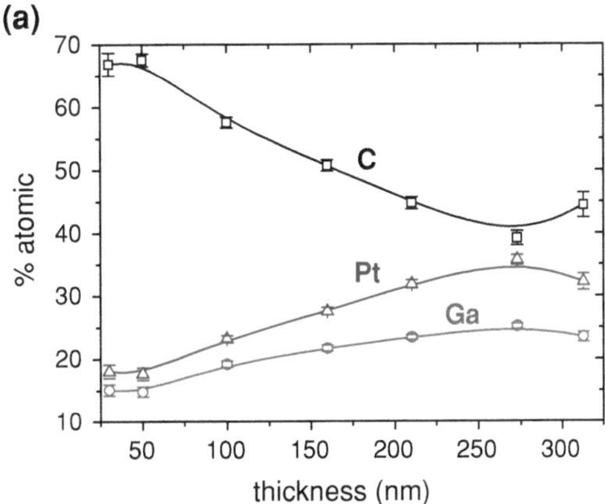

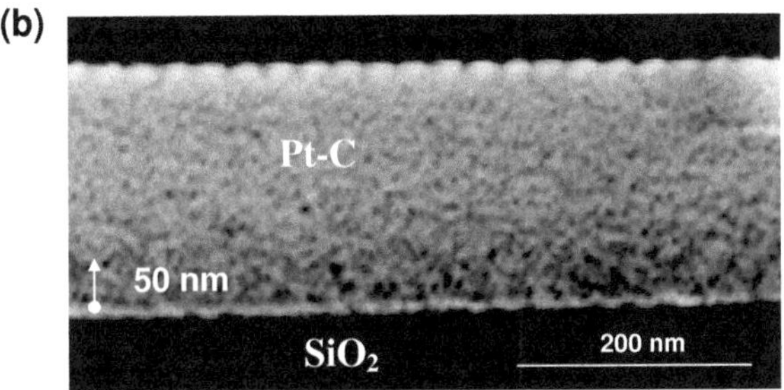

Fig. 5.2 a Atomic percentage compositions of Pt–C NWs as a function of their thickness, determined by EDX. We observe a clear difference in composition for the NWs smaller than 50 nm in comparison with the others, where a higher metallic content is present. Oxygen is always below 1%. **b** STEM *dark-field image* of a 200 nm-thick lamella prepared from a NW. Approximately, the first deposited 50 nm have a *darker* contrast in comparison with *upper layers*, indicating a higher carbon concentration at the initial stages of growth

range [20], because of the large energy gap present in SiO_2. The SE are considered to be the main cause of the dissociation of the precursor gas molecules [21, 22]. Thus, at the beginning of the growth process, a smaller amount of SE is emitted in comparison with the subsequent stages, when the effective substrate becomes the initial C–Pt–Ga deposit. This argument is in agreement with the penetration range of Ga at 30 kV in a SiO_2 and Pt–C substrate (of the order of 30–50 nm), as SRIM calculations indicate [23]. A second explanation can be related to heating effects present during the deposit, which is a crucial point in a FIBID and FEBID process [21, 24, 25]. As the structure grows in height the heat flow would pass from a 3–dimensional regime to a pseudo 2–dimensional regime. Thus, the higher Pt percentage for the top of the structure could be associated with heat dissipation that is less effective than at the beginning of the growth [21, 25]. The gradient in metal–carbon concentration with thickness gives rise to a broad phenomenology in the transport properties of the NWs, as will be described later in this work.

5.1.3.2 XPS Measurements

In order to gain a more detailed insight into the nature of the FIBID-Pt, we have performed a depth profile XPS analysis in a micrometric sample, which remained in contact with the atmosphere for 1 day after deposition. Survey spectra show the

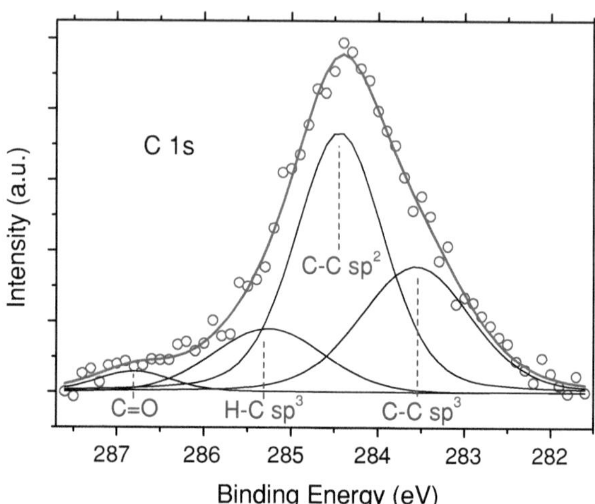

Fig. 5.3 C 1s XPS core level for the 170 nm-thick layer (corresponding to a 30 nm in-depth etching). The *solid lines* are the components in which the spectrum is decomposed. The resulting fit is superimposed on the experimental data (*open circles*). The energies found for the peaks are associated with C–C sp^3, C–C sp^2, C–H sp^3 and C=O (see text for details). The carbon present in the deposit has approximately 50% sp^2 hybridization through all the deposit thickness

presence of carbon, platinum, gallium and oxygen, with concentrations similar to those obtained by EDX. A detailed analysis was done by studying the evolution of the absorption edges: C 1s, Pt 4f, Ga 3d and O 1s, as a function of thickness.

(i)   Analysis of the C 1s edge:

The C 1 s spectra have a constant profile through all the thickness. In Fig. 5.3, we can see the particular case for thickness = 170 nm (30 nm ion-etched), where four chemical components are used to fit the spectrum. These components are, in order of binding energy: sp^3 carbon with carbon–carbon bonds (C–C sp^3, 283.6 ± 0.1 eV) [26], sp^2 carbon with carbon–carbon bonds (C–C sp^2, 284.3 ±0.1 eV) [26, 27], sp^3 carbon with hydrogen-carbon bonds (C–H sp^3, 285.3 ± 0.1 eV) [26, 27], and a C=O contribution, at 286.2 ± 0.1 eV [28]. This last minor peak is compatible with the O 1s spectra (not shown here), where a peak is found at 531.8 ± 0.1 eV. All the peaks have a FWHM of 1.3 ± 0.3 eV [27]. Thus, we find that the carbon present in the deposit has a sp^2 hybridization of around 55%. This percentage is slightly lower than in previous results obtained by Energy Electron Loss Spectroscopy in suspended Pt nanostructures grown by FEBID, where an approximately 80% ratio has been reported [29].

(ii) Ga 3d$^{5/2}$ edge (not shown here):

A peak is present at 18.7 ± 0.1 eV, corresponding to metallic gallium [30].

(iii) Analysis of the Pt 4f$^{7/2}$ edge:

In Fig. 5.4 we show the Pt 4f depth profile. Paying attention to the 4f$^{7/2}$ edge, all the spectra can be fitted by one single peak, associated to metallic Pt. This is true in all cases except for the top layer (200 nm), which is much broader, where an extra peak is necessary to fit the spectrum. This extra contribution is located at

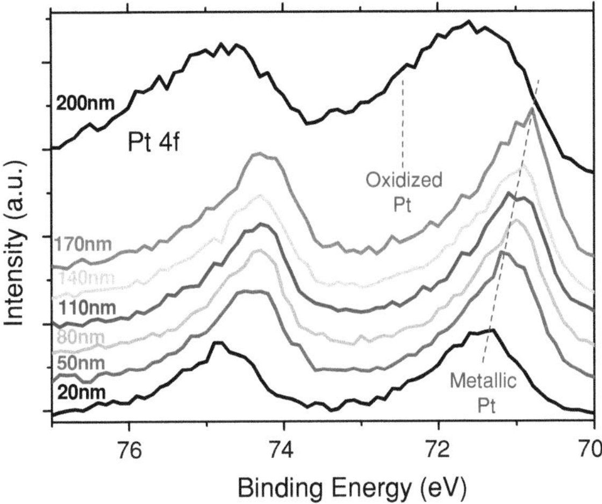

Fig. 5.4 Pt 4f core level XPS depth profile. Successive spectra were obtained after a 30 nm argon etching process. The *numbers* in the figure refer to the remaining sample thickness. The external surface of the deposit presents a broader spectrum than the rest, as a consequence of an extra peak present, associated with platinum oxide or hydroxide (*left dashed line*). For the rest of surfaces, the spectra can be fitted by only one contribution, associated to metallic platinum. This peak presents a shift towards lower binding energies as the probed layer is increasingly internal (oblique, *right dashed line*), as well as a decrease in the asymmetry of the Pt 4f peak. We correlate this evolution with thickness with a progressive increase of the cluster sizes as the deposit gets thicker

72.5 ± 0.1 eV, indicating the presence of oxidized Pt in the external surface, in the form of an oxide or hydroxide form [28]. In the O 1s edge, a peak at 531.2 ± 0.1 eV gives account of it. This oxidation seems to be of outstanding relevance for the transport properties of the NWs with small thickness (lower than 50 nm) when exposed to ambient conditions (see Sect. 5.1.3.4).

For the rest of Pt 4f spectra, we can see substantial differences with thickness. First, a progressive shift towards higher BE occurs from the more external part of the deposit (170 nm) to the more internal one (20 nm) (see oblique line). For the 170 nm spectrum, the maximum is located at 70.7 ± 0.1 eV. This BE matches the range typically found at the surface of Pt crystals [31]. For the 20 nm spectrum, the maximum shifts in BE to the value of 71.2 ± 0.1 eV. The second important thickness-dependent feature is associated with the asymmetry of the peaks. The asymmetry in XPS peaks for metallic systems is caused by the screening of the core–hole by low energy electron–hole excitation at the valence band [19]. By fitting the peaks with the line shape proposed by Doniach and Sunjic [18] to describe this effect (Eqs. 5.1.a, 5.1.b), we find that whereas the outer surfaces present an asymmetry factor $\alpha = 0.18 \pm 0.04$, matching perfectly with the value for bulk Pt [19], the more internal surfaces have $\alpha = 0.10 \pm 0.06$. This lower value for the asymmetry is presumably caused by a decrease in the local density of d-states at the Fermi level [32, 33]. The lifetime width of the core hole created by photoemission is almost constant through the entire thickness, $\gamma = 0.56 \pm 0.05$. This value is higher than for bulk Pt, which is typical for small metal clusters [32, 34]. The two features found as the deposited thickness increases (enhancement in the degree of asymmetry and shift towards lower BE) would be compatible with an increase in size of the Pt clusters [32, 34] with thickness.

5.1.3.3 "In Situ" Measurements of the Resistance as the NWs are Grown

As we explained in previous sections, we have measured the resistance of the NWs as the growth process is performed. In Fig. 5.5 we show how the resistivity (ρ) of a NW evolves as a function of deposition time in a typical experiment. The results were reproduced over several experiments.

The highest value for ρ corresponds to ~ 1 GΩ in resistance, and to ~ 1 kΩ for the lowest value. We have correlated the process time (T_p) with the deposited thickness by doing cross-section inspections of NWs created at different times. Thus, we can also express the results in terms of resistivity, by using the relation:

$$\rho = R\frac{w \times t}{L} \tag{5.2}$$

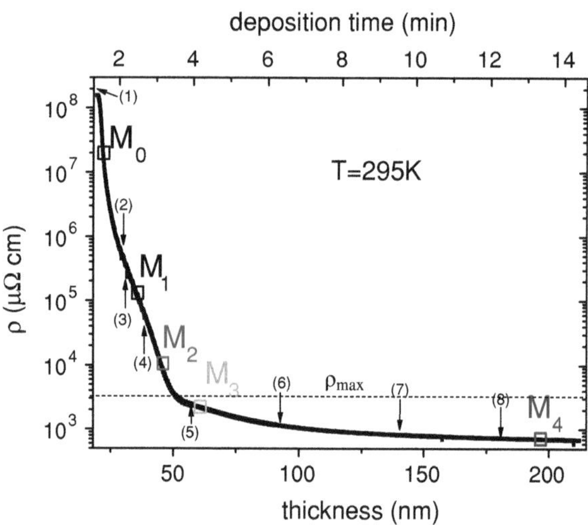

Fig. 5.5 Resistivity as a function of deposition time and thickness for a typical process. The resistivity varies by more than four orders of magnitude as a consequence of a change in composition. A thickness of 50 nm marks the transition between a non-metallic conduction (t < 50 nm) and metallic one (t > 50 nm). The resistivity saturates to a constant value for t > 150 nm, $\rho \approx 700$ $\mu\Omega$ cm (about 65 times higher than bulk Pt). The maximum resistivity for metallic conduction in non-crystalline materials (ρ_{max}), as calculated by Mott, is marked. This line separates the insulator NWs from the metallic ones. The *numbers* in *parentheses* correspond to different probe currents during the measurements. (*1*): 5 nA, (*2*): 10 nA, (*3*): 30 nA, (*4*): 100 nA, (*5*): 1 μA, (*6*): 3 μA, (*7*): 5 μA, (*8*): 10 μA

We can see that under these conditions the resistivity of the NWs starts from a value above 10^8 $\mu\Omega$ cm. An abrupt decrease is produced as the thickness increases, lowering its resistivity in 4 orders of magnitude when the thickness reaches 50 nm. From that moment on, ρ decreases slightly, saturating to a constant value of ~ 700 $\mu\Omega$ cm (around 65 times higher than the value for bulk Pt) for $t > 150$ nm. The negligible thickness dependence of ρ for thickness >50 nm was previously evidenced in reference [8]. We have not observed any increase in ρ for long T_p as a consequence of the disorder created by ions, such as that reported for W and Pd-FIBID [13]. This procedure has been repeated many times (~ 10), the behavior being the same in all cases. From these experiments, we conclude that the resistivity at room temperature is highly dependent on the NW thickness. This result explains the large diversity of values for ρ_{RT} of FIBID-Pt NWs reported in the literature [1–12]. We must emphasize that the chosen parameters are of great importance in the particular values of the resistivity for a given thickness, since effects such as heating, diffusion, interaction of the beam with substrate, etc., can differ depending on the conditions. We are not able to reach resistivities as low as the ones reported by Lin et al. [3–5]. The different precursor gas used for deposition, as it was pointed out in Sect. 5.1.1., seems to be the most probable reason for it. The resistivity as a function of thickness crosses the value reported by Mott for the maximum metallic resistivity for a non-crystalline material, $\rho_{max} \approx 3000$ $\mu\Omega$ cm [35] at ≈ 50 nm (see dashed line in Fig. 5.5). Taking into account this criterion, we would roughly expect a non-metallic conduction for samples with $\rho > \rho_{max}$, and a metallic conduction for $\rho < \rho_{max}$.

To study the nature of the conductivity in deposits having different thickness, we have fabricated several NWs, by means of stopping the growth process when a determined value of the resistivity is reached. Hereafter these NWs will be generically referred to as: M_0 ($\sim 10^7$ $\mu\Omega$ cm), M_1 ($\sim 10^5$ $\mu\Omega$ cm), M_2 ($\sim 10^4$ $\mu\Omega$ cm), M_3 ($\sim 2 \times 10^3$ $\mu\Omega$ cm) and M_4 (~ 700 $\mu\Omega$ cm), see squares in Fig. 5.5. The resistance in all cases is stable under the high vacuum atmosphere, once the deposit has stopped. Current versus voltage (I–V) measurements have been performed "in situ", at room temperature. In Fig. 5.6 we show examples of the differential conductance (G = dI/dV, obtained numerically) as a function of voltage. For M_4 (Fig. 5.6e) and M_3 (Fig. 5.6d), I–V is linear (G constant), so a metallic character is inferred. However, for M_2 (Fig. 5.6c), G(V) is slightly parabolic. This result agrees perfectly with reference 209, where this behavior is observed for a deposit of $\rho = 3078$ $\mu\Omega$ cm. The same non-linear effect is even much stronger for M_1 (Fig. 5.6b). A different behavior in G(V) occurs for the most resistive sample, M_0 (Fig. 5.6a), where a peak in the differential conductance is found at low bias, decreasing from ~ 10 $n\Omega^{-1}$ to a constant value of 4 $n\Omega^{-1}$ for voltages higher than 0.5 V. G(V) measurements confirm Mott's criterion for the maximum resistivity for metallic conduction in non-crystalline materials [35]. We discuss in detail the evolution of G(V) with thickness in Sect. 5.1.4.

We can understand these strong differences for the electrical transport properties at room temperature taking into account the results obtained in Sect. 5.1.3.1., where a gradient in composition is observed for the NWs as a function of

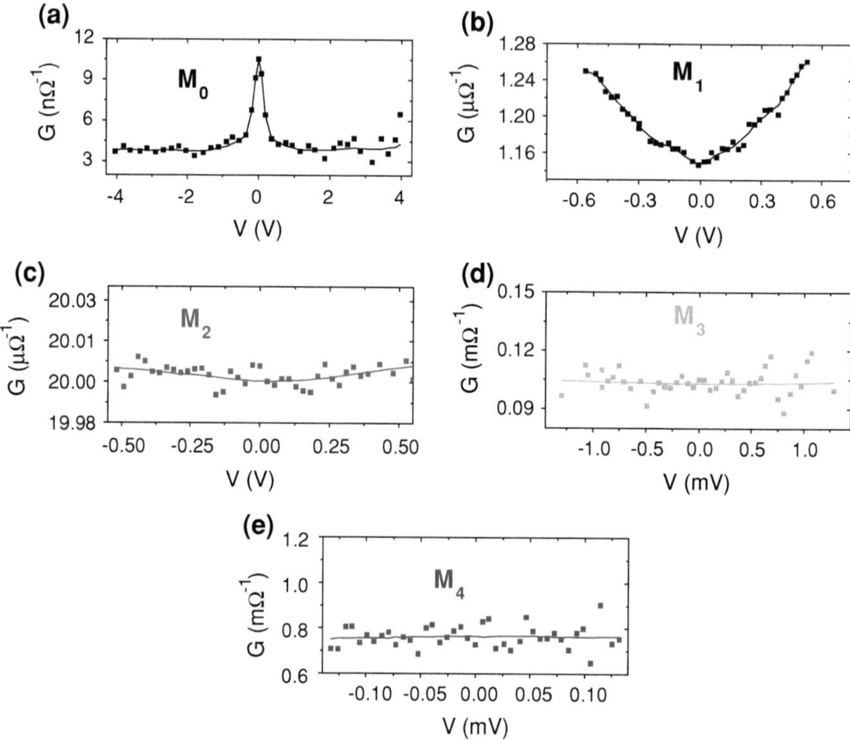

Fig. 5.6 Differential conductivity versus voltage for samples in different regimes of conduction, marked in the $\rho(t)$ curve (Fig. 5.5). G(V) has been obtained by a numerical differentiation of current versus voltage curves. In M_0 an exponential decrease in the differential conductance is present. In M_1 a positive tendency for the G(V) curve is observed, being hard to see for M_2. In M_3 and M_4 G(V) is constant (see text for details)

thickness. For NWs thinner than approximately 50 nm, the metal (Pt + Ga) content is lower than 33% (atomic), resulting in resistivity values orders of magnitude higher than those for pure Pt, together with non-linear features in the I–V curves. For NWs thicker than 50 nm, the metal content is higher than 33%. In this case, the resistivity reaches a minimum value (about 65 times higher than for Pt), and metallic I–V characteristics are observed. A more detailed discussion will be presented in Sect. 5.1.4.

5.1.3.4 Temperature Dependence of the Electrical Properties

For a better understanding of the transport mechanisms in the Pt–C NWs, we have measured the thermal dependence of the resistance, as well as I–V curves at different temperatures. We have studied NWs in the representative ranges shown in Sect. 5.1.3.3.: M_0, M_1, M_2, M_3 and M_4. The first substantial difference between samples is related to the modification of the resistance when the NWs come into contact with ambient atmosphere. For M_4, M_3 and M_2, we did not observe important changes in R, whereas for M_1 the resistance was roughly doubled, and for M_0 it became higher than 1 GΩ. These changes are clearly associated with the oxidation effect for Pt at the top part of the deposits after exposure to ambient conditions (see Sect. 5.1.3.2.).

In Fig. 5.7a we show the resistivity versus temperature [$\rho(T)$] curves for M_1, M_2, M_3 and M_4). We can see a radically different behavior depending on the resistivity at room temperature. In the case of M_1, a highly negative thermal dependence is found, increasing more than two orders of magnitude when $T = 25$ K (RRR $= \rho_{RT}/\rho_{25K} \approx 9 \times 10^{-3}$). Nevertheless, in the rest of the samples (see their evolution with temperature in detail in Fig. 5.7b) a minor dependence on temperature is found, with RRR slightly lower than 1 (≈ 0.7 for M_2, ≈ 0.8 for M_3, ≈ 0.9 for M_4). This is a clear evidence of the important role played by the disorder created by the Ga ion irradiation during the growth, as well as by the inhomogeneous nature of the deposit. We must remark again that a positive

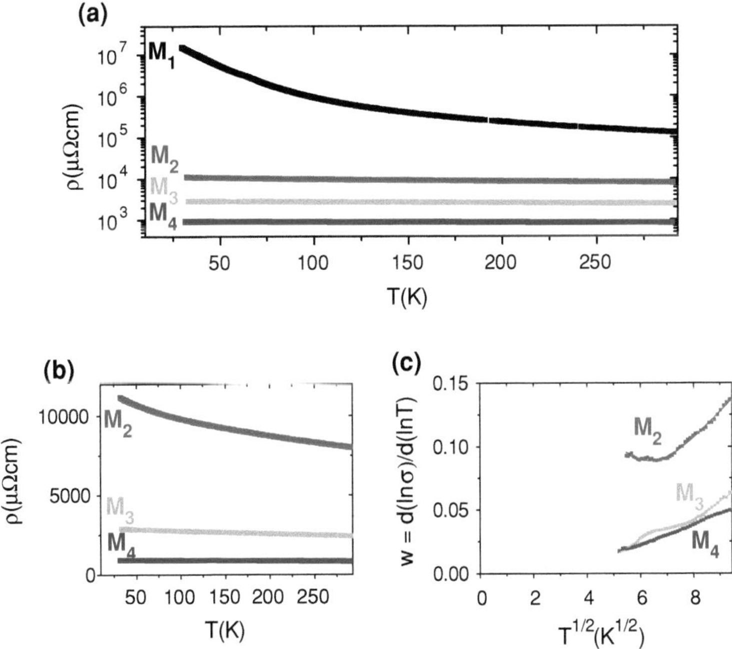

Fig. 5.7 **a** Resistivity as a function of temperature for NWs in four selected regimes of resistivity (note the logarithmic scale). The thermal dependence is radically different depending on the value of ρ at room temperature, indicating a metal–insulator transition as a function of thickness (therefore as a function of composition). **b** $\rho(T)$ curves of M_2, M_3 and M_4 in normal scale, to observe in detail the variations in resistivity for these samples. **c** w (Eq. 5.3 in text) as a function of $T^{1/2}$. From this analysis, together with the I–V behavior, we deduce that for NWs in the regime of M_3 and M_4 the conduction is metallic, being insulator for the rest of the selected regimes

thermal coefficient for FIBID-Pt has been shown in literature [3–5], but using a different gas precursor.

We have used the $\rho(T)$ measurements to determine if the NWs series presents a metal–insulator transition (MIT) with thickness, by calculating the variable:

$$w(T) = \frac{d(\ln \rho^{-1})}{d(\ln T)} \tag{5.3}$$

The dependence of w upon T is far more sensitive for determining the transition than $\rho(T)$, as has been stressed by Möbius et al. [36]. $w(T \to 0) = 0$ implies a metallic character of the conduction, whereas either a diverging or finite value for $w(T \to 0)$ indicates that the sample is an insulator. In our case, the lowest temperature attainable is too high (25 K) to apply this criterion strictly. However, in M_2, as shown in Fig. 5.7c, $w(T)$ is far from tending to zero, whereas in M_3 and M_4 the behavior of $w(T)$ has this tendency. We could then infer that M_3 and M_4 have metallic conduction, being insulators for the rest. This is in perfect agreement with the results obtained at room temperature (see Sect. 5.1.3.3) and, as will be shown below, with the I–V study done as a function of temperature. Therefore, the results support a scenario where the Pt–C NWs grown by FIB present a MIT as a function of thickness (metal/carbon composition). If a low metal content is present, the conduction is attributable to carrier hopping into localized states, whereas a high enough amount of metallic inclusions guarantees that transport will take place by carriers in extended states. This explains the diversity of results found in literature for the dependence of Pt-FIBID resistivity with temperature. It should be noticed that amorphous carbon by itself is well known to present a MIT by tuning the sp^2–sp^3 ratio [37, 38], but in our case the amount of metal present in the deposits is the key parameter for determining the mechanism responsible for conduction.

We have studied in detail $\rho(T)$ in the non-metallic sample M_1. By fitting the curve using an activation dependence:

$$\rho(T) = \rho_\infty \exp^{\left(\frac{T_0}{T}\right)^N} \tag{5.4}$$

it is found that the best exponent N is 0.5. Variable-range hopping (VRH) of electrons between localized states, as predicted by Mott, has the same functional dependence, but with $N = 0.25$ [35]. Efros and Shklovskii demonstrated that a VRH process when the Coulomb interaction between sites is taken into account (ES-VRH) results in a Coulomb gap, yielding an exponent of 0.5, instead of 0.25 [39]. There exists a critical temperature T_c above which Mott-VRH is fulfilled, whereas below T_c, Coulomb interactions become important, and the conduction is by ES-VRH. As discussed in reference [40] for FEBID-Pt nanostructures, a T_c higher than room temperature is expected if reasonable assumptions are made for the values of two unknown quantities in this material: the density of states at the Fermi energy, and the dielectric constant. The observed thermal dependence of M_1 suggests that the same arguments are valid in this case. We should also mention that $\rho(T)$ in M_2 also follows this dependence in the low T-regime.

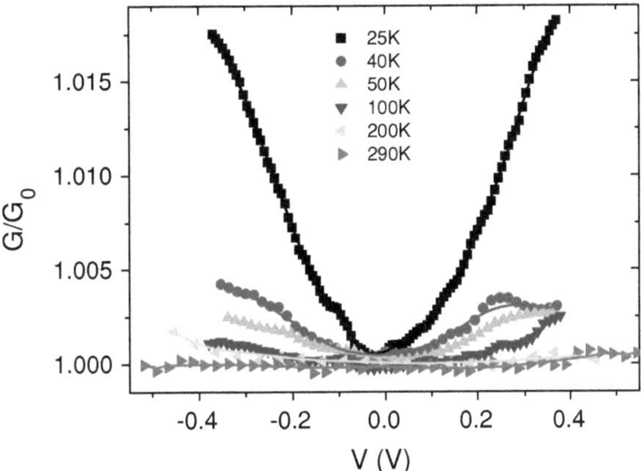

Fig. 5.8 Normalized differential conductance (G/G_0) as a function of voltage for M_2. A positive differential conductance is observed, increasing the non-linearity as T is lowered

We have also performed I–V curves as a function of temperature for all the samples and, as in the case of measurements inside the chamber, differentiated numerically the curves to obtain the differential conductance G. First, for M_4 and M_3, I–V curves are linear for all temperatures (not shown here), indicating a metallic character in all the range of T studied. This is not the case for M_2 and M_1. The results are in perfect agreement with the w(T) analysis done before.

In the case of M_2, we show in Fig. 5.8 the normalized conductance G/G_0 versus voltage for selected temperatures, where G_0 is the lowest value for G. The non-linearity effects increase as T diminishes, reaching 2% [$100 \times (G/G_0 - 1)$] at 25 K. We can see that the dependence for this sample is the same as that for the one reported in reference 209, where the increase in the differential conductance with voltage was understood in terms of the Glazman-Matveev model [41] for multi-step tunnelling occurring between the Pt-Ga nanocrystals embedded in the C-insulator matrix. We will discuss this dependence in Sect. 5.1.4.

In Fig. 5.9 we show the normalized conductance for M_1. For temperatures higher than 100 K, the same dependence as in M_2 is exhibited. This sample is much more insulating than M_2, so the non-linearity at 100 K, for instance, is around 5%. A richer behavior is found for temperatures below 100 K. As can be observed, a peak in G(V) appears at zero bias, as a consequence of a decrease of the conductance for low voltages. It is the same behavior found for the sample M_0 for "in situ" measurements, and we will show in Sect. 5.2 that this behavior is also present in NWs grown by FEBID. For higher V the same non-linear dependence for G(V) as in sample M_2 is observed. We can see in detail the thermal evolution of the peak in G(V) as a function of temperature in the inset of Fig. 5.9. The height of the peak increases in magnitude as the temperature is lowered, reaching its maximum at 60 K (around 80% at V = 0). For lower T, the peak becomes broader and smaller in height. On the other hand, the non-linear effect at high voltages (increase of the differential con-ductance) is above 150% at 30 K. We interpret in detail all these results in the following section.

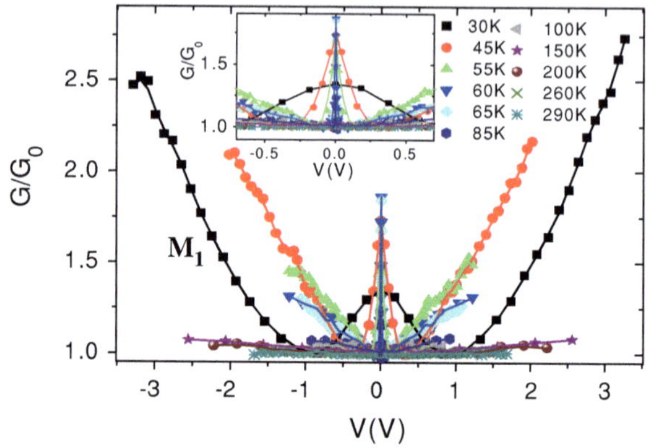

Fig. 5.9 Normalized differential conductance (G/G_0) as a function of voltage for M_2. The positive differential conductance, also observed in M_2, is present in M_1 for high bias. Superimposed to this effect, a decrease in the differential conductance occurs for T < 100 K. The evolution of this phenomenon is shown in detail in the *inset*

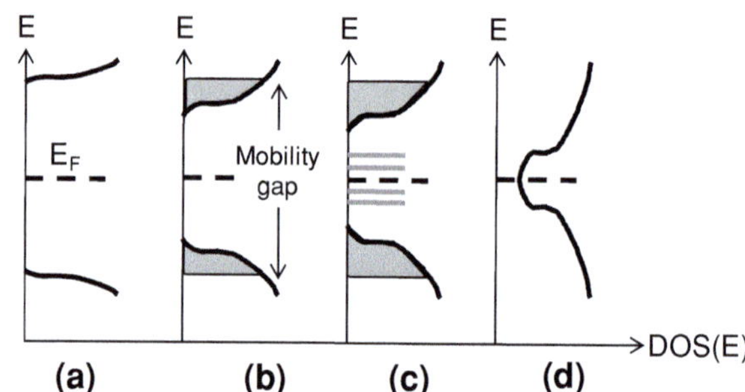

Fig. 5.10 Density of states D(E) for **a** a semiconductor **b** a disordered semiconductor, such as a-C **c** disordered semiconductor with metallic inclusions **d** a disordered metal. The *grey* states are localized, whereas white are extended

5.1.4 Discussion of the Results

As has been shown in previous sections, the methodology we have used to study the resistivity of Pt–C NWs created by FIB, using "in situ" measurement of the resistance while the deposit is being performed, has allowed us to determine substantial differences in the electrical transport properties with thickness, i.e., with composition. In this section we discuss the different mechanisms for conduction existing in the NWs, and their temperature dependence, which depend on their composition.

The MIT we observe in the NWs as a function of thickness can be understood under the Mott-Anderson theory for non-crystalline materials [35] (see a scheme of the evolution of the DOS of the material with metal doping in Fig. 5.10). Amorphous carbon (a-C) is a disordered semiconductor. The gap when the hybridization is partially sp^2, as we have determined by the XPS analysis, is typically around 1–2.5 eV [42, 43]. Because of the intrinsic disorder present in the nanodeposits, band tails appear inside the gap as localized states, with the so called mobility edge separating localized and non-localized states inside the band. If inside a matrix composed by a-C, metallic inclusions are introduced in a low percentage (M_0 and M_1), this will mainly result in the incorporation of localized defects within the band gap. The conductivity, especially at low temperatures, will be due to hopping conduction between these defects. As the deposit induced by ions continues (M_2), two effects occur: an increase of disorder favored by the continuous ion irradiation, and the incorporation of a higher percentage of metal, as we showed by EDX characterization. The disorder will result in a decrease of the gap by an enlargement of the band tails, resulting in a partial delocalization [42]. Therefore, the hopping will not be only between the impurity states within the gap, but also by means of the localized states inside the band, below the mobility edge. An even higher percentage of metallic inclusions will create a continuum of levels, inducing a transition to the metallic regime when the concentration is above the percolation edge (M_3 and M_4). Under this scenario, we can understand the different results reported in literature regarding the transport properties of Pt–C deposits grown by FIB (Table 5.1). Depending on the carbon concentration, the conduction will be metallic, by means of extended states, or insulating, by a hopping process between localized states.

To understand the mechanisms for conduction in the insulating regime we now focus our discussion on the voltage dependence of the differential conductance for the samples M_0, M_1 and M_2 (Figs. 5.6a, 5.8 and 5.9). We will first interpret the $G(V)$ behavior for the less resistive sample, M_2 (Fig. 5.8). As we previously commented, the thermal dependence at low temperatures is associated with an ES-VRH. In this theory, above a critical electric field E^*, the hopping is field-dominated, and the conductance is expressed by [44]:

$$G = G(0) \exp\left(-\sqrt{\frac{E^*}{E}}\right) \qquad (5.5.a)$$

$$E^* \approx \frac{1.4e}{4\pi\varepsilon_0\varepsilon\xi_L^2} \qquad (5.5.b)$$

where E is the applied electric field, ε is the dielectric constant, and ξ_L is the localization length. We show in Fig. 5.11a a typical fit performed for M_2 at 25 K. In this case, the data fit quite well to the dependence of (5.5.a) for fields above 100 V/cm. From (5.5.b), and assuming a dielectric constant $\varepsilon \approx 4$ for an a-C rich Pt NW [40, 45], we can estimate the localization length as a function of temperature, for T < 125 K (Fig. 5.12). Thus, $\xi_L(M_2) \approx 80$ nm for this range of temperatures.

The same treatment has been done for the differential conductance of M_1 (Fig. 5.11b), but in this case the ES-VRH dependence in fulfilled in the whole range of temperatures measured. The fit has been done to the part of $G(E)$ where the conductance increases with the electric field. An example for T = 45 K is shown in Fig. 5.11b, where the positive exponential dependence predicted by (5.5.a) is fitted to the experimental data for fields E > 350 V/cm. The localization length derived for the fits is shown in Fig. 5.12. $\xi_L \approx 30\text{–}50$ nm for high temperatures, decreasing to a value of about 20 nm for low temperatures. The lowest temperature attained in the setup is quite high (25 K), and could explain why ξ_L is

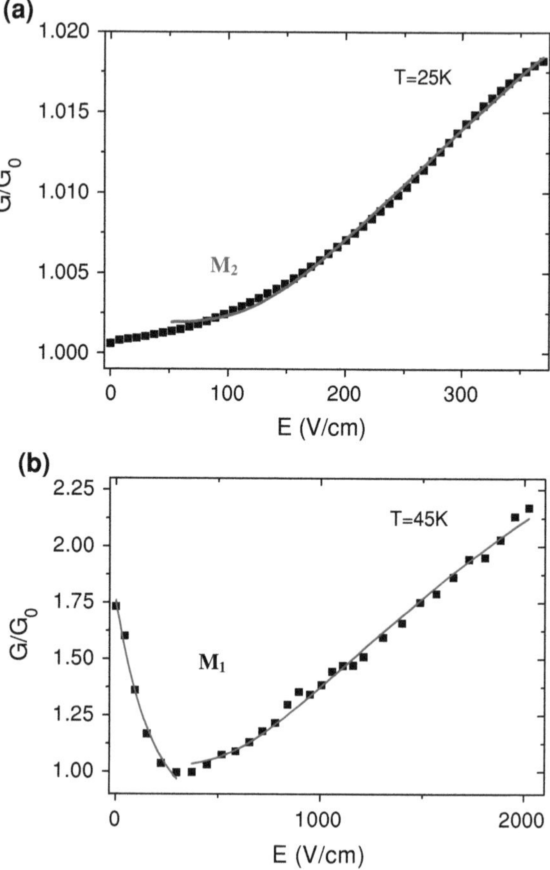

Fig. 5.11 Normalized differential conductance (G/G_0) as a function of the electric field, for M_2 at T = 25 K **a** and M_1 at T = 45 K **b**. *Solid lines* are fits to the experimental data. In **a**, the fit is done for E > 50 V/cm to the ES-VRH model for $E > E^*$ (see Eqs. 5.5.a and 5.5.b in text). In **b**, the same model is used for the fit if E > 350 V/cm. If E < 350 V/cm, the Shklovskii theory to describe an exponential decrease in the hopping conductivity is used (Eq. 5.6.a and 5.6.b in text)

significantly bigger than the size of the metallic crystals in the carbon matrix (around 3 nm, see Sect. 5.2.2.4.), in terms of a thermal effect.

We should now discuss the origin of the low-bias decrease of the differential conductance found for the most resistive sample in the series (M_0) at room temperature, as well as for M_1 for T < 100 K. Bötger and Bryksin [46] and Shklovskii and co-workers [47, 48] developed analytical models for the hopping conduction in disordered systems under strong electric fields. Those models predict that in certain conditions an exponential decrease of the hopping conductivity with the electric field occurs, even producing a negative differential conductance in certain cases. Experimentally, this was demonstrated in lightly doped and weakly compensated silicon [48, 49]. Recently, numerical simulations confirm that this behavior of the conductance as a function of the applied voltage is inherent to the hopping transport in this conduction regime [50]. The theory developed predicts an exponential decrease of the conductance with the electric field in the following form:

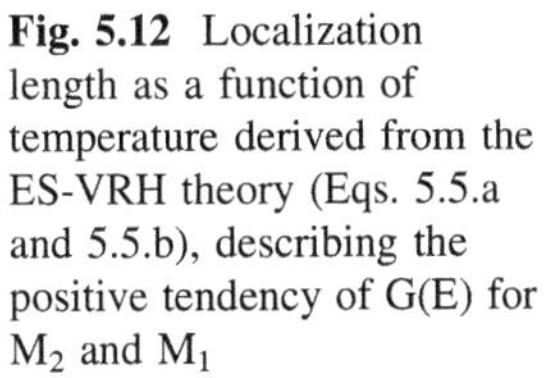

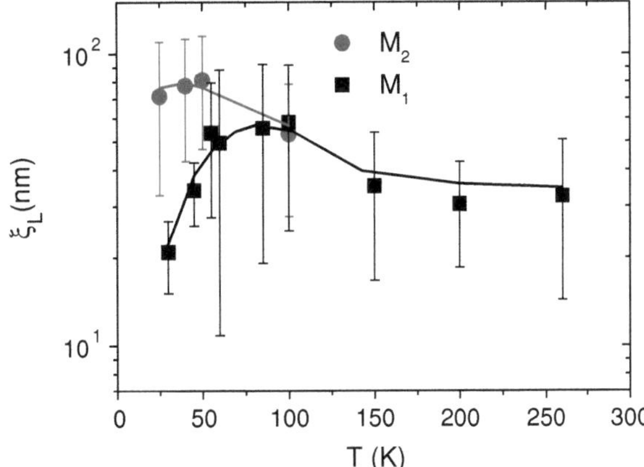

Fig. 5.12 Localization length as a function of temperature derived from the ES-VRH theory (Eqs. 5.5.a and 5.5.b), describing the positive tendency of G(E) for M_2 and M_1

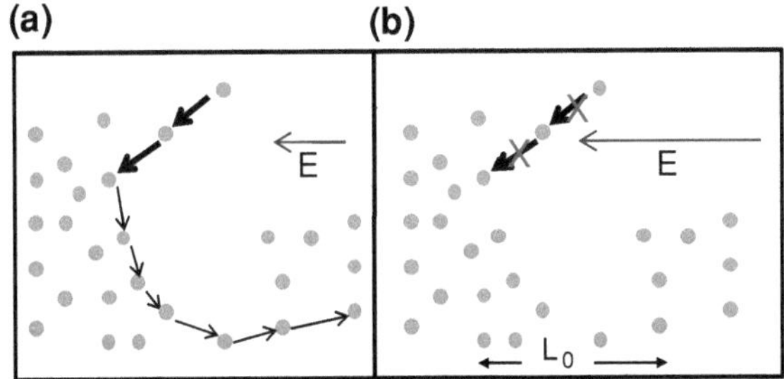

Fig. 5.13 Hopping conductivity of electrons in a cluster of impurities (*Circles*). **a** At moderate electric fields, the electrons located in a dead end can "escape" making backwards movements. **b** At very high fields, the ends act as traps for electrons, since backwards movements are not possible, producing the decrease of conductance with E

$$G(E) \approx G(0) \exp\left(\frac{-eEL_0}{2kT}\right) \tag{5.6.a}$$

$$L_0 = \frac{R}{3}\left(\frac{1.74R}{\xi_L}\right)^{0.88} \tag{5.6.b}$$

where $R = N^{-1/3}$ is the mean inter-impurity distance, ξ_L is the localization length, and L_0 is the typical length of the region where the electron gets "trapped". This equation is valid if $eEL_0 >> kT$. To understand the essence of the effect, let us think of a cluster composed of impurities, responsible for the hopping conduction (see Fig. 5.13). In a hopping process at moderate fields, the electron hops from one site to other, opposite to the electric field. In this process, there exists the possibility that the electron arrives at a "dead-end", i.e., an impurity centre whose neighboring sites are distant enough to make it energetically unfeasible that the electron to continue its way. If this happens, the electron tries to move in the direction of the field to escape from this "trap" to further continue hopping again opposite to E. However, if the conduction occurs under a strong electric field, some of these backward movements are forbidden. Thus, electrons get trapped in such dead–ends, and consequently the current density decreases with E. For a detailed description of the phenomenon see references [46–50]. We associate the decrease in G(V) found in M_0 and M_1 to this effect, in complete agreement with the previous interpretations done for other ranges of resistivities and fields. As observed experimentally in Figs. 5.6a and 5.9, after the drop in G(V) at low bias, a change to positive slopes appears at even higher fields, which is a direct consequence of the high value of the field, which provides enough energy to overcome the traps in the system, resulting in the dependence of VRH with field described by Eqs. 5.5 [49].

We have used the Eqs. 5.6 to fit the experimental G(E) data of M_0 and M_1. In Fig. 5.11b, the example for M_1 at 45 K is shown, where the fit is done for $0 < E < 350$ V/cm. For $T > 50$ K, the criterion $eEL_0 \gg kT$ is not fulfilled, resulting in unphysical values for L_0 (≈ 10 μm), and consequently for the localization length (below the nanometer). We therefore centre our study for $T < 50$ K, where $L_0 \approx 100$ nm. If we assume an impurity concentration similar to that which exists in silicon when the exponential decrease in G(V) is observed, $N \approx 10^{16}$ cm^{-3} [48, 49], an estimate of the ξ_L can be made. For low temperatures, $\xi_L \approx 10$ nm. This value approaches the one estimated by the ES-VRH fits for the higher-bias increase of G(E), of around 20 nm (Fig. 5.12). The non-perfect agreement between both values seems logical, since assumptions regarding the dielectric constant and the concentration of impurities are done in both models. Nevertheless, the values are in the same order of magnitude, evidencing the consistency of the analysis performed. If we apply the same procedure for M_0 at room temperature, we obtain $L_0 \approx 200$ nm, resulting in a localization length $\xi_L \approx 6$ nm if an impurity concentration $N = 5 \times 10^{15}$ cm^{-3} is assumed.

5.2 Comparison of NWs Created by FEBID and FIBID

The aim of this section is to compare the electrical conduction properties of Pt–C NWs grown by FEBID and by FIBID. The deposits are compared for thicknesses high enough so that FIBID-NWs have metallic conduction (in the range of M_4).

5.2.1 Experimental Details

We detail the parameters different from those exposed in Sect. 5.1.2.

5.2.1.1 Deposition Parameters

The deposition parameters for FEBID-NWs are: $(CH_3)_3Pt(CpCH_3)$ precursor gas, $(T_G) = 45°C$ precursor temperature, $(V_{FEB}) = 10$ kV beam voltage, $(I_{FEB}) = 0.54$ nA beam current, substrate temperature $= 22°C$, dwell time $= 1$ µs, chamber base pressure $= 10^{-7}$ mbar, process pressure $= 3 \times 10^{-6}$ mbar, beam overlap $= 0\%$, distance between GIS needle and substrate $(L_D) = 1.35$ mm. For FIBID-NWs, the parameters are the same as in Sect. 5.1.2.1., except for $T_G = 45°C$, and $L_D = 1.35$ mm.

5.2.1.2 "In Situ" Electrical Measurements

The procedure is identical to that described in Sect. 5.1.2.2., except that the resistance-versus-time measurements were done in a two-probe configuration, by connecting the conductive microprobes via a feedthrough to a Keithley 2000 multimeter located out of the dual beam chamber, which allows measuring the deposit resistance by two-probe measurements provided that its value is below 120 MΩ. A constant current of value 700 nA was applied. The lead resistance, only about 13 Ω, guarantees no significant influence on the measured resistance of the deposit, which is at least two orders of magnitude larger as will be shown later.

5.2.1.3 HRTEM Analysis

The high-resolution TEM study was carried out using a Jeol 2010F equipment$^{®}$ operated at 200 kV (point to point resolution 0.19 nm) on Pt deposits grown on Cu TEM grids with a supporting carbon membrane.

5.2.2 Results

5.2.2.1 EDX

EDX analysis reveals a lower metallic/carbon content in the case of FEBID deposits with respect to the FIBID ones (in atomic percentage):

- FEBID-NWs: C (%) = 87.8 ± 0.2, Pt (%) = 12.2 ± 0.2
- FIBID-NWs: C (%) = 71.8 ± 0.2, Pt (%) = 17.4 ± 0.2, Ga (%) = 10.8 ± 0.2).

5.2.2.2 "In Situ" Measurements of the Resistance as the NWs are Grown

In Fig. 5.14 we compare Pt–C NWs by FEBID and FIBID with similar thickness in order to give evidence for the strong difference in the resistance of the deposits. Three FEBID and FIBID deposits are shown to illustrate the reproducibility of the

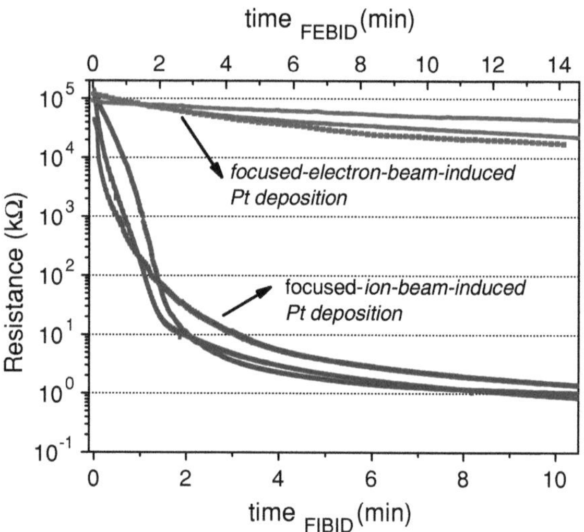

Fig. 5.14 Resistance versus time measurements, comparing several Pt NWs grown by FEBID and FIBID. The result gives account of the big differences in the conduction mechanism for both types of deposits, as a consequence of the distinct types of interactions in both cases. For the sake of clarity, zero time is the starting point of the resistance monitorization in all cases

results. In the case of FEBID (FIBID) deposits, after 6.3 (1.5) min, the resistance reaches 120 MΩ and we start the resistance monitorization. After additional deposition time of 14.6 (10.7) min, the final resistance is found to be 23.5 MΩ (1 kΩ). Even though the final thickness of all the deposits is roughly the same ($\sim$160 nm), measured by conventional cross-section inspection, the resistance is four orders of magnitude lower for FIBID-Pt than for FEBID-Pt. Thus, in the case of Pt by FEBID the resistivity is about 10^7 $\mu\Omega$ cm, whereas it is about 800 $\mu\Omega$ cm cm in the case of Pt by FIBID, in any case much higher than for bulk Pt, of 10.8 $\mu\Omega$ cm. The higher metallic doping in the case of FIB-deposits is responsible for the important differences in resistivity (see Sect. 5.2.2.1).

We must notice that in spite of slightly different parameters for the growth of FIBID-NWs than those selected in Sect. 5.1, the final resistivity is approximately the same. Thus, after the profuse discussion of the previous section, we conclude that the FIBID-Pt NW used for comparison with those deposited by electrons, are in the metallic regime of conduction (samples in the range of M_4).

In Fig. 5.15, still in situ, we compare the four-probe current-versus-voltage dependence after finishing the growth of one FEBID and one FIBID Pt deposit of similar thickness to those of Fig. 5.14. Whereas the FIBID deposit shows the expected linear dependence, the FEBID deposit, with $\rho \approx 10^7$ $\mu\Omega$ cm shows a non-linear dependence. The differences between both types of behavior lie on the different metallic content of the nanodeposits. The linear behavior of the FIBID-Pt is the expected response for a metallic or quasi-metallic system. On the other hand, the non-linear behavior observed in the FEBID-Pt deposit can be explained on the basis of the relevant role played by the semiconducting carbonaceous matrix. Interestingly, as the behavior for the conductance of the FEBID-Pt wire decreases up to $\approx$0.6 V and then starts to increase for higher voltages (see inset in Fig. 5.15b). This dependence is qualitatively the same as the most resistive

Fig. 5.15 In-situ four-probe current-versus-voltage measurements of FEBID and FIBID Pt deposits with thickness ~ 160 nm. The measurements in **a** FIBID-Pt NW indicate linear dependence between current and applied voltage whereas those in **b** FEBID-Pt NW indicate non-linear dependence. The *inset* shows the anomalous dependence of conductance versus voltage, understood by the trapping of carriers in sites under strong electric fields

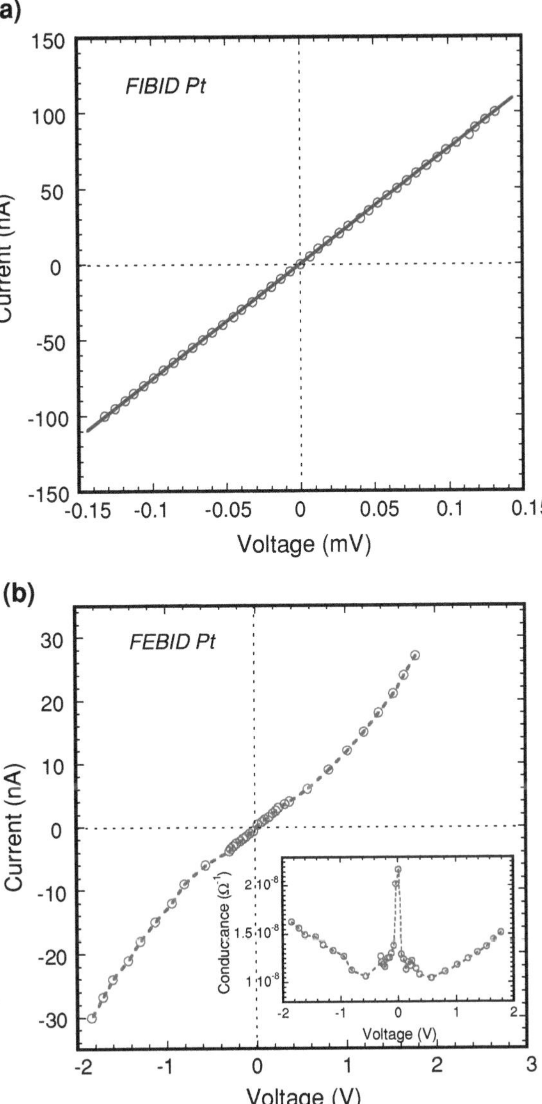

FIBID-NWs studied in Sect. 5.1 (M_0). The high electric fields produce the trapping of electrons, inducing a decrease in the conductance for low bias [46–50] (see Sect. 5.1.4 for details).

5.2.2.3 Temperature Dependence of the Electrical Properties

In Fig. 5.16 the dependence of the resistance with temperature for the FEBID deposit is shown (the behavior of the FIBID-NW is almost negligible, as M_4 in

Fig. 5.16 Temperature variation of the resistance of a FEBID-Pt NW, showing a semiconducting dependence. The inset shows the analysis of data in the form of ln(G) versus $T^{-1/2}$. An ES-VRH is deduced from this fit, for $T > 200$ K

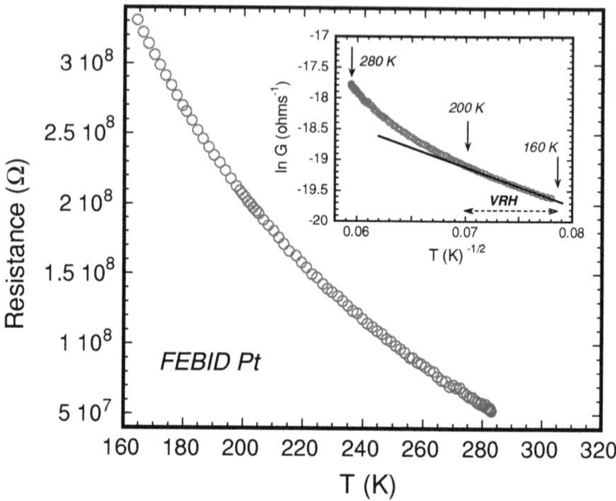

Sect. 5.1). R(T) shows a strong semiconductor behavior, and at $T = 150$ K the resistance is of the order of 1 GΩ, the maximum resistance that our experimental setup can measure (a factor of 5.5 compared to 300 K). In the inset of Fig. 5.16, the representation of ln(G) versus $T^{-1/2}$ indicates that below 200 K the conduction mechanism fits the ES-VRH [39] (Eq. 5.4, with N = 0.5). The same temperature dependence has been previously found in FEBID-Pt deposits by Tsukatani et al. [10]. Above 200 K, the thermal energy is high enough to allow other electronic processes to be involved such as hopping via thermal activation [35], leading to a deviation from the $T^{-1/2}$ law.

As a result of this comparison, and taking into account the conclusions of Sect. 5.1.4:

(i) The mechanisms for conduction of Pt by FEBID and FIBID for thicknesses roughly higher than 100 nm are different, with a metallic conduction for deposits induced by ions, and by an activation mechanism for those induced by electrons. By changing process parameters such as the beam energy or current, within accessible ranges, we do not find significant differences in the transport results, FEBID-Pt always being several orders of magnitude more resistive than the FIBID one.

(ii) It is evident the big similarities between FEBID deposits and highly resistive FIBID deposits, in spite of the different mechanisms for deposition, the presence of Ga ions for FIBID, and the increase of disorder by the interaction of ions. As in the study performed in this section, previous reports comparing both materials mainly focus on the differences found between low-ρ-FIBID-Pt and FEBID-Pt [1, 51]. Further work is necessary to study the similarities and differences between both systems in the range of resistivities where the coincident features appear.

5.2.2.4 HRTEM Images

For completeness, we include HRTEM images of both deposits (This piece of work forms part of the thesis of Rosa Córdoba (University of Zaragoza: rocorcas@unizar.es)) that illustrate the reason for the different mechanisms of conduction.

As can be observed in Fig. 5.17, the FEBID and FIBID Pt–C wires consist of ellipsoidal crystalline Pt grains embedded in an amorphous carbonaceous matrix. The crystalline Pt grains are easily identified due to the clear observation of Pt atomic planes with the expected planar distances. Thus, fast-Fourier transforms of the images give as a result the presence of diffraction spots corresponding to the planar distances of 0.2263, 0.1960, 0.1386, and 0.1132 nm, which correspond respectively to the (111), (200), (202) and (222) atomic planes of FCC Pt, with

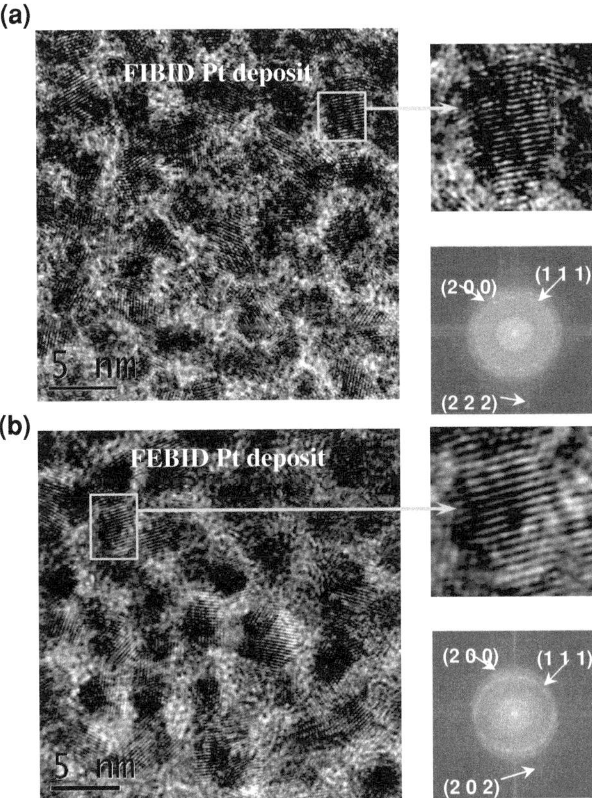

Fig. 5.17 HRTEM images of a Pt nanodeposit by FIBID **a** and FEBID **b** in both cases grown at 30 kV beam energy on top of a TEM Cu grid covered with a thin supporting holey carbon membrane. One Pt grain has been selected in each case for magnification and clear observation of the corresponding atomic planes. The Fast-Fourier-Transform of the full image gives diffraction spots that correspond to the (200), (111), (222) and (202) atomic planes of FCC Pt

lattice parameter of 0.3924 nm. An average size of Pt crystallites has been obtained from the measurement of about 50 individual grains in each image through the counting of the number of atomic planes. It is found that no matter the used beam energy and current, in all the FEBID and FIBID deposits the average Pt crystallite size is about 3.2 ± 0.8 nm. Thus, from these HRTEM images, the microstructure of the FEBID and FIBID Pt deposits is similar. The higher percentage of Pt in deposits by FIB brings about the percolation of grains, inducing metallic characteristics in the electrical transport.

5.3 Conclusions

We have performed a detailed study of the transport properties of Pt–C NWs grown by Focused-Ion/Electron-Beam induced deposition. By controlling the resistance as the deposit is done, we can see the different regimes for the resistivity existing in this material.

In the case of FIBID, we have deposited NWs with five different thicknesses, that is, in five different regimes of (over 5 orders of magnitude) conductivity (A similar work to ours has been recently reported in FEBID-W by Huth et al., New J. of Phys. 11, 033032 (2009). However, they do not observe a MIT). By studying NW resistivity as a function of temperature, as well as the dependence with voltage on the differential conductivity, we observe a metal–insulator transition in the NWs as a function of thickness (carbon–metal content). This transition is interpreted under the theory of Mott-Anderson for non-metallic disordered materials. As the NW increases in height, a higher concentration of metal is introduced in the semiconducting carbon matrix. This gives rise to the appearance of impurity levels in the band gap, together with the localization of states in the formed band tails. As the growth continues, the introduction of metal clusters finally results in a delocalization of the electron wave-functions, and NWs above 50 nm present metallic conduction.

This work explains the discrepancy existing for previous Pt–C nanostructures created by FIB, where in some cases a metallic character was reported, in contrast to others, where insulating conduction was observed. It also gives a counterpoint to the traditional work when growing metal nanostructures by FIB, since we demonstrate the possibility to deposit materials with different conduction characteristics, making it, in principle, feasible to fabricate insulator or metal NWs with the same technique and precursor, just by controlling and changing the growth parameters. This is of special relevance because of its potential extrapolation to other materials grown by FIBID and FEBID, since most of the gas precursors used by these techniques use organo-metallic gases, as it is the case here.

The comparison of metallic-FIBID and FEBID wires can be explained by the different composition of both types of deposits, as a consequence of the different mechanisms responsible for deposition, with a lower metallic doping for deposits induced by focused electrons. The common features between FEBID-NWs and low metallic FIBID-NWs are also evidenced by electrical, spectroscopic and imaging measurements.

The discovery of an exponential decrease in the differential conductance behavior in low-metallic doped carbon NWs constitutes an experimental evidence in a nanometric structure of the validity of the theories developed for hopping conduction in strong fields. The existence of this trend in both types of deposits (grown by FEB and FIB) indicates that this conduction process naturally appears in Pt–C NWs grown with focused-beam methods.

References

1. R.M. Langford, T.-X. Wang, D. Ozkaya, Reducing the resistivity of electron and ion beam assisted deposited Pt. Microelectron. Eng. **84**, 748 (2007)
2. L. Peñate-Quesada, J. Mitra, P. Dawson, Non-linear electronic transport in Pt nanowires deposited by focused ion beam. Nanotechnology **18**, 215203 (2007)
3. J.-F. Lin, J.P. Bird, L. Rotkina, P.A. Bennett, Classical and quantum transport in focused-ion-beam-deposited Pt nanointerconnects. Appl. Phys. Lett. **82**, 802 (2003)
4. J.-F. Lin, J.P. Bird, L. Rotkina, A. Sergeev, V. Mitin, Large effects due to electron–phonon-impurity interference in the resistivity of Pt/C-Ga composite nanowires. Appl. Phys. Lett. **84**, 3828 (2004)
5. J.-F. Lin, L. Rotkina, J.P. Bird, A. Sergeev, V. Mitin, Transport in Pt nanowires fabricated by focused-ion-beam induced deposition and electron-beam induced deposition. IPAP Conf. Ser. **5**, 17 (2004)
6. T. Tao, J. Ro, J. Melngailis, Focused ion beam induced deposition of platinum. J. Vac. Sci Technol. B **8**, 1826 (1990)
7. J. Puretz, L.W. Sawson, Focused ion beam deposition of Pt containing films. J. Vac. Sci Technol. B **10**, 2695 (1992)
8. K.A. Telari, B.R. Rogers, H. Fang, L. Shen, R.A. Weller, D.N. Braski, Characterization of platinum films deposited by focused ion beam-assisted chemical vapor deposition. J. Vac. Sci Technol. B **20**(2), 590 (2002)
9. J.M. De Teresa, R. Córdoba, A. Fernández-Pacheco, O. Montero, P. Strichovanec, M.R. Ibarra, Origin of the Difference in the Resistivity of As-Grown Focused-Ion- and Focused-Electron-Beam-Induced Pt Nanodeposits. J. Nanomat. **2009**, 936863 (2009)
10. Y. Tsukatani, N. Yamasaki, K. Murakami, F. Wakaya, abd.M. Takai, Transport properties of Pt nanowires grown by beam-induced deposition. Jpn. J. Appl. Phys. **44**, 5683 (2005)
11. K. Dovidenko, J. Rullan, R. Moore, K.A. Dunn, R.E. Geer, F. Heuchling, FIB-assisted Pt deposition for carbon nanotube integration and 3-D nanoengineering. Mater. Res. Soc. Symp. Proc. **739**, H7.7.1 (2003)
12. G. De Marzi, D. Lacopino, A.J. Quinn, G. Redmond, Probing intrinsic transport properties of single metal nanowires: direct-write contact formation using a focused ion beam. J. Appl. Phys. **96**, 3458 (2004)
13. D. Spoddig, K. Schindler, P. Rodiger, J. Barzola-Quiquia, K. Fritsch, H. Mulders, P. Esquinazi, Transport properties and growth parameters of PdC and WC nanowires prepared in a dual-beam microscope. Nanotechnology **18**, 495202 (2007)
14. M. Prestigiacomo, L. Roussel, A. Houël, P. Sudraud, F. Bedu, D. Tonneau, V. Safarov, H. Dallaporta, Studies of structures elaborated by focused ion beam induced deposition. Microelectron. Eng. **76**, 175 (2004)
15. F. Hernández-Ramírez, A. Tarancón, O. Casals, E. Pellicer, J. Rodríguez, A. Romano-Rodríguez, J.R. Morante, S. Barth, S. Mathur, Electrical properties of individual tin oxide nanowires contacted to platinum electrodes. Phys. Rev. B **76**, 085429 (2007)

16. T. Scwamb, B.R. Burg, N.C. Schirmer, D. Poulikakos, On the effect of the electrical contact resistance in nanodevices. Appl. Phys. Lett. **92**, 243106 (2008)
17. G. Boero, I. Utke, T. Bret, N. Quack, M. Todorova, S. Mouaziz, P. Kejik, J. Brugger, R.S. Popovic, P. Hoffmann, Submicrometer Hall devices fabricated by focused electron-beam-induced deposition. Appl. Phys. Lett. **86**, 042503 (2005)
18. S. Doniach, M. Sunjic, Many-electron singularity in X-ray photoemission and X-ray line spectra from metals. J. Phys. C **3**, 285 (1970)
19. S. Hüfner, G.K. Wertheim, Core-line asymmetries in the x-ray-photoemission spectra of metals. Phys. Rev. B **11**, 678 (1975)
20. K. Ohya, T. Ishitani, Monte Carlo study of secondary electron emission from SiO_2 induced by focused gallium ion beams. Appl. Surf. Sci. **237**, 606 (2004)
21. I. Utke, P. Hoffmann, J. Melngailis, Gas-assisted focused electron beam and ion beam processing and fabrication. J. Vac. Sci. Technol. B **26**, 1197 (2008)
22. A.D. Dubner, A. Wagner, J. Melngailis, C.V. Thompson, The role of the ion-solid interaction in ion-beam-induced deposition of gold. J. Appl. Phys. **70**, 665 (1991)
23. SRIM is a group of programs which calculate the stopping and range of ions into matter using an ion-atom collisions treatment. The simulator is available in http://www.srim.org/#SRIM
24. W.F. van Dorp, C.W. Hagen, A critical literature review of focused electron beam induced deposition. J. Appl. Phys. **104**, 081301 (2008)
25. I. Utke, J. Michler, P. Gasser, C. Santschi, D. Laub, M. Cantoni, P.A. Buffat, C. Jiao, P. Hoffmann, Cross section investigations of compositions and sub-structures of tips obtained by focused electron beam induced deposition. Adv. Eng. Mater. **7**, 323 (2005)
26. S. Takabayashi, K. Okamoto, K. Shimada, K. Motomitsu, H. Motoyama, T. Nakatani, H. Sakaue, H. Suzuki, T. Takahagi, Chemical structural analysis of diamond like carbon films with different electrical resistivities by X-ray photoelectron spectroscopy. Jpn. J. Appl. Phys. **47**, 3376 (2008)
27. J. Díaz, G. Paolicelli, S. Ferrer, F. Comin, Separation of the sp^3 and sp^2 components in the C1s photoemission spectra of amorphous carbon films. Phys. Rev. B **54**, 8064 (1996)
28. V. Alderucci, L. Pino, P.L. Antonucci, W. Roh, J. Cho, H. Kim, D.L. Cocke, V. Antonucci, XPS study of surface oxidation of carbon-supported Pt catalysts. Mater. Chem. Phys. **41**, 9 (1995)
29. S. Frabboni, G.C. Gazzadi, A. Spessot, TEM study of annealed Pt nanostructures grown by electron beam-induced deposition. Physica E **37**, 265 (2007)
30. C.D. Wagner, W.M. Riggs, L.E. Davis, J.F. Moulder, G.E. Muilenberg, *Handbook of X-Ray Photoelectron Spectroscopy* (Perkin-Elmer Corporation, Physical Electronics Division, Eden Prairie, MI, 1979)
31. R.C. Baetzold, G. Apai, E. Shustorovich, R. Jaeger, Surface core-level shifts for Pt single-crystal surfaces. Phys. Rev. B **26**, 4022 (1982)
32. T.T.P. Cheung, X-ray photoemission of small platinum and palladium clusters. Surf. Sci. **140**, 151 (1984)
33. A.Y. Stakheev, Y.M. Shulga, N.A. Gaidai, N.S. Telegina, O.P. Tkachenko, L.M. Kustov, K.M. Minachev, "New evidence for the electronic nature of the strong metal-support interaction effect over a Pt/TiO_2 hydrogenation catalyst. Mendeleev. Commun. **11**(5), 186 (2001)
34. K. Fauth, N. Schneider, M. Heßler, G. Schütz, Photoelectron spectroscopy on Pt atoms and clusters deposited on C(0001). Eur. Phys. J. D **29**, 57 (2004)
35. N.F. Mott, E.A. Davis, in *Electronic Processes in Non-Crystalline Materials* (Oxford University Press, Oxford, 1971)
36. A. Möbius, C. Frenzel, R. Thielsch, R. Rosenbaum, C.J. Adkins, M. Cheiber, H.-D. Bauer, R. Grötzschel, V. Hoffmann, T. Krieg, N. Matz, H. Vinzelberg, M. Witcomb, Metal-insulator transition in amorphous $Si_{1-x}Ni_x$: evidence for Mott's minimum metallic conductivity. Phys. Rev. B **60**, 14209 (1999)
37. V. Prasad, Magnetotransport in the amorphous carbon films near the metal-insulator transition. Solid State Commun. **145**, 186 (2008)

38. S. Bhattacharyya, S.R.P. Silva, Transport properties of low-dimensional amorphous carbon films. Thin Solid Films **482**, 94 (2005)
39. A.L. Efros, B.I. Shklovskii, Coulomb gap and low temperature conductivity of disordered systems. J. Phys. C **78**, L49 (1975)
40. Z.-M. Liao, J. Xun, D.-P. Yu, Electron transport in an array of platinum quantum dots. Phys. Lett. A. **345**, 386 (2005)
41. L.I. Glazman, K.A. Matveev, Inelastic tunneling across thin amorphous films. Sov. Phys. JETP **67**, 1276 (1988)
42. R.U.A. Khan, J.D. Carey, R.P. Silva, B.J. Jones, R.C. Barklie, Electron delocalization in amorphous carbon by ion implantation. Phys. Rev. B **63**, 121201(R) (2001)
43. J. Robertson, Diamond-like amorphous carbon. Mater. Sci. Eng. R **129**, 129 (2002)
44. D. Yu, C. Wan, B. Wherenberg, P. Guyot Sionnest, Variable range hopping conduction in semiconductor nanocrystal solids. Phys. Rev. Lett. **92**, 216802 (2004)
45. A. Grill, Amorphous carbon based materials as the interconnect dielectric in ULSI chips. Diam. Relat. Mater. **10**, 234 (2001)
46. H. Böttger, V.V. Bryksin, Effective medium theory for the hopping conductivity in high electric fields. Phys. Status Solidi B **96**, 219 (1979)
47. N. van Lien, B.I. Shklovskii, Hopping conduction in strong electric fields and directed percolation. Solid State Commun. **38**, 99 (1981)
48. D.I. Aladashvili, Z.A. Adamia, K.G. Lavdovskii, E.I. Levin, B.I. Shklovskii, Negative differential resistance in the hopping-conductivity region in silicon. JETP **47**, 466 (1988)
49. A.P. Mel'nikov, Y.A. Gurvich, L.N. Shestakov, E.M. Gershenzon, Magnetic field effects on the nonohmic impurity conduction of uncompensated crystalline silicon. JETP Lett **73**, 50 (2001)
50. A.V. Nenashev, F. Jansson, S.D. Baranovskii, R. Österbacka, A.V. Dvurechenskii, F. Gebhard, Hopping conduction in strong electric fields: negative differential conductivity. Phys. Rev. B **78**, 165207 (2008)
51. S. Lipp, L. Frey, C. Lehrer, E. Demm, S. Pauthner, H. Rysell, A comparison of focused ion beam and electron beam induced deposition processes. Microelectron. Reliab. **36**, 1779 (1996)

Chapter 6
Superconductor W-based Nanowires Created by FIBID

In this chapter the results obtained for W-based nanowires grown by Focused-Ion-Beam-Induced-Deposition will be shown. This material is a superconductor for temperatures below ~ 5 K.

The first part of the chapter shows the characterization of the microstructure and chemical composition, by High-Resolution-Transmission-Electron-Microscopy and X-ray Photoelectron Spectroscopy studies. The second part consists of the study of the electrical measurements of the wires, studying if the superconducting properties are preserved for wire width in the nanoscale. Finally, and for completeness, spectacular measurements by Scanning-Tunneling-Spectroscopy will be shown, as a result of collaboration with the group of Prof. Vieira and Dr. Suderow (UAM, Madrid).

6.1 Introduction

6.1.1 Previous Results in FIBID-W

Sadki et al. [1, 2] discovered in 2004 that W-based nanodeposits created by 30 kV-FIBID, using $W(CO)_6$ as gas precursor, were SCs, with a critical temperature of around 5 K, substantially higher than in metallic crystalline W ($T_c = 0.01$ K) [3]. This high T_c was associated with the amorphous character of the material, as determined by X-ray diffraction measurements, as well as by the presence of C and Ga inside the deposits (in atomic percentage, determined by EDX: W/C/Ga = 40/40/20%). Values for the critical field ($Bc_2 = 9.5$ T) and current density ($J_c = 0.15$ MA/cm^2) were deduced from electrical measurements in wires of 10 μm (length) $\times$ 300 nm (width) $\times$ 120 nm (thickness).

A few more results can be found in the literature after this important result. The most relevant references are:

A. Fernandez-Pacheco, *Studies of Nanoconstrictions, Nanowires and Fe₃O₄ Thin Films*, Springer Theses, DOI: 10.1007/978-3-642-15801-8_6,
© Springer-Verlag Berlin Heidelberg 2011

1. Luxmoore et al. [4]: analogous W-based deposits by electrons (FEBID) do not present SC transition down to 1.6 K, suggesting the importance of Ga in the SC properties in FIBID-W, by becoming implanted and creating amorphization.
2. Spoddig et al. [5]: measurement of resistance during the growth of 500 nm-wide W-C wires, and study of Bc_2 and J_c, with similar results to those of [1].
3. Li et al. [6]: T_c can be tuned from 5.0 to 6.2 K by varying the FIB current. This is explained by the modulation of disorder created by ions near the metal–insulator transition, in accordance with the theory developed by Osofsky et al. [7].
4. Kasumov et al. [8, 9]: an interesting application of these W-nanodeposits is their use as resistance-free leads, for the electrical characterization of contacted single organometallic molecules.

6.1.2 Transition Superconductor Temperatures in W Species

Table 6.1 shows different types of tungsten and tungsten species, and their associated superconducting transition temperatures. We see that only amorphous W, in combination with other elements, presents transition temperatures similar to that for FIBID-W, $T_c \approx 5$ K.

6.2 Experimental Details

The FIBID superconducting nanodeposits were fabricated at room temperature in the dual-beam equipment detailed in Sect. 2.1.2, which uses a 5–30 kV Ga^+ FIB. The $W(CO)_6$ precursor gas is brought onto the substrate surface by a GIS, where it becomes decomposed by the FIB. Common parameters for this deposition process are: precursor gas temperature $= 50°C$, vol/dose $= 8.3 \times 10^{-2}$ $\mu m^3/nC$, dwell time $= 200$ ns, beam overlap $= 0\%$, chamber base pressure $= 10^{-7}$ mbar, process pressure $= 6 \times 10^{-6}$ mbar, distance between GIS needle and substrate $= 150$ μm.

Table 6.1 Different W and W- species found in nature, together with their SC transition temperatures

Material	T_c (K)	Reference
Metallic W	0.01	Gibson and Hein [3]
β-W	3.2–3.35	Bond et al. [10]
(Hexagonal)-WC	–	Willens and Buehler [11]
(FCC)-WC	10	Willens and Buehler [11]
W_2C	2.7	Willens and Buehler [11]
Evaporated amorphous W + Re	3.5–5	Collver and Hammond [12]
Sputtered amorphous W + Si, Ge (20–40%)	4.5–5	Kondo [13]
CVD amorphous W + C (40%)	4	Miki et al. [14]
FIBID-W	4.1–6	[1, 2, 4–6, 8]

6.2.1 HRTEM Analysis

The HRTEM study was carried out using a FEG-TEM Jeol 1010 equipment® operated at 200 kV (point to point resolution 0.19 nm) on 20 nm-thick W nanodeposits directly grown on TEM grids for electron transparency in the experiment.

6.2.2 XPS Measurements

A 100 µm × 100 µm × 100 nm sample was deposited on a Si wafer, using $V_{BEAM} = 30$ kV, $I_{BEAM} = 5$ nA.

The measurements were done in the equipment detailed in Sect. 2.3.2. A 5 kV-argon etching, with current densities of the order of 0.15 A/m^2 was done, to obtain a depth profile of the deposits (probing every 20 nm). For the quantitative analysis of the XPS spectra, we used pseudo-Voigt peak profiles with a 10–20% Lorentzian contribution, subtracting a Shirley-type background.

6.2.3 Electrical Measurement in Rectangular (Micro- and Nano-) Wires

Rectangular wires were deposited on previously micropatterned $SiO_2//Si$, where Al or Ti pads were evaporated after an optical lithography process, to produce contact pads for the magnetotransport measurements (see details of lithography in Sect. 2.1.2). The dimensions of the wires are: length = 12 µm, variable width and thickness (two types). The first type consists of microwires (M_i) with width ≈ 2 µm and thickness ≈ 1 µm, and the second type consists of nanowires (N_i) with widths in the range 100–250 nm and thickness of 100–150 nm.

The transport measurements (up to 9 T magnetic fields) were made in a PPMS system® (Quantum Design).

6.3 HRTEM Analysis of FIBID-W

In Fig. 6.1, the HRTEM images of one selected FIBID-W nanodeposit (under 30 keV) are shown [this piece of work forms part of the thesis of Rosa Córdoba 15]. The microstructure is globally homogeneous at the scale of tens of nanometers. At the scale of about 1 nm, some contrast appears in the images likely due to local differences in the relative amount of the heavier W compared to the lighter C and Ga. No crystalline structure is observed anywhere by direct inspection of the images. This is confirmed by the Fast Fourier Transform of the image, shown in the inset, which does not present any diffraction spot or ring. This is in contrast for example to FIBID-Pt, where metallic grains are embedded in an a-C matrix (see

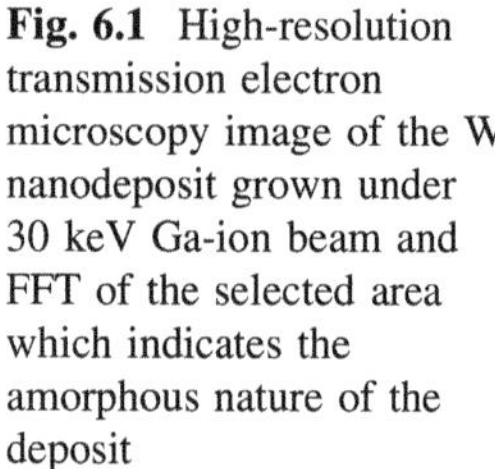

Fig. 6.1 High-resolution transmission electron microscopy image of the W nanodeposit grown under 30 keV Ga-ion beam and FFT of the selected area which indicates the amorphous nature of the deposit

Sect. 5.2.2.4), but in good agreement with previous XRD [1] and HRTEM [4] experiments in FIBID-W nanodeposits. From these measurements we can infer the amorphous nature of the FIBID-W nanodeposits.

6.4 XPS Study of FIBID-W

XPS measurements were performed to determine the chemical composition and the chemical state of the deposits.

In Fig. 6.2, the in-depth study (every 20 nm) shows a quite homogeneous deposit. The composition, in atomic percentage, is: $W = (39.8 \pm 7.6)\%$, $C = (43.2 \pm 3.7)\%$, $Ga = (9.5 \pm 2.8)\%$, $O = (7.5 \pm 1.5)\%$. Similar results were obtained with standard EDX characterization inside the chamber, and are in perfect agreement with [1].

In Fig. 6.3, typical fits for the W 4f and C 1s spectra are shown.

1. W 4f: The main peak for the $W4f^{7/2}$ appears at 31.4 eV, associated with metallic W [16, 17]. A minor peak at 32.1 eV, corresponding to tungsten carbide (WC) [16, 17] is also present. A separation of 2.1 eV between the $4f^{7/2}$ and the $4f^{5/2}$ components is found in all cases. The ratio between both species through the thickness is $W^0/WC = 5.17 \pm 0.53$. In Table 6.2, details of the fit are provided.
2. C 1s: The spectrum is significantly broad, and can be deconvoluted in three peaks (Table 6.3). One at 283.1 eV, due to WC [18, 19], and the others at 284.3 and 285.2 eV, which correspond to amorphous carbon, sp^2 and sp^3 atoms,

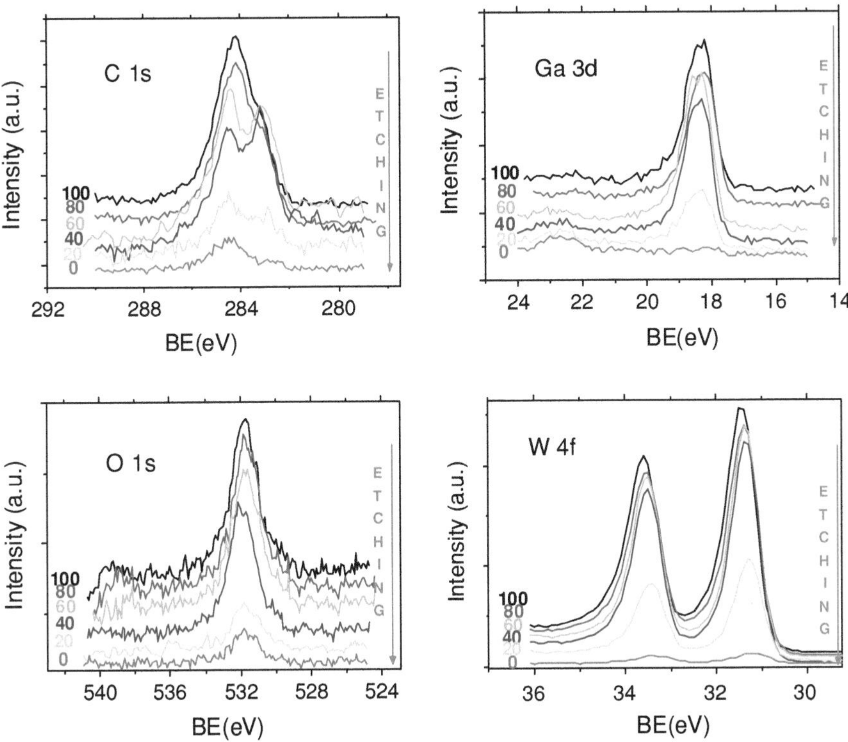

Fig. 6.2 XPS depth profile analysis of a FIBID-W microdeposit. The surface is probed every 20 nm by Ar etching (see *vertical arrow* on the right of the graphs). The composition is homogeneous along the thickness, with the deposit composed of C, W, Ga and O. The numbers in every figure refer to the remaining sample thickness in nanometers

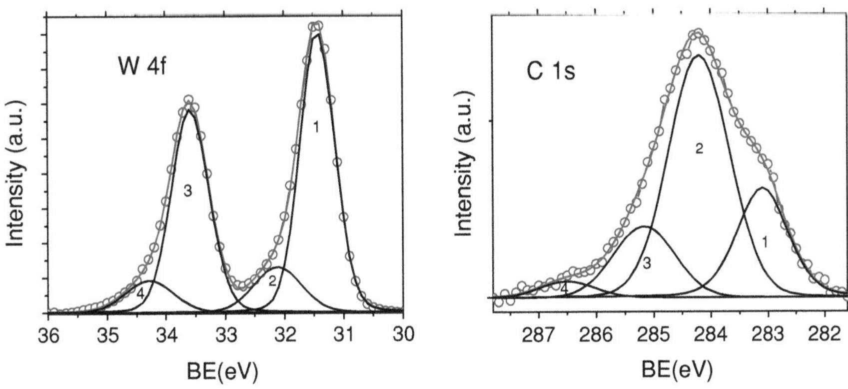

Fig. 6.3 Typical fits for W 4f (left) and C1s (right) spectra along the depth profile. Together with the experimental data (*dots*), the fit is superimposed (*line*), result of the convolution of the different components (*lines*). The details of the fit are shown in Tables 6.2 and 6.3

Table 6.2 W 4f components

Peak	BE (eV)	FWHM (eV)	% Lorentzian	Species
1	31.4	0.68	10	W ($4f^{7/2}$)
2	32.1	1.03	10	WC ($4f^{7/2}$)
3	33.6	0.68	10	W ($4f^{5/2}$)
4	34.3	1.03	10	WC ($4f^{5/2}$)

The peak numbers appear in Fig. 6.3. The FWHM of the same species was fixed at the same value (same mean life-time). The area between components was constrained to the relationship 4:3, taking into account the degeneracy of each level

Table 6.3 C 1s components

Peak	BE (eV)	FWHM (eV)	% Lorentzian	Species
1	283.1	1.06	20	WC
2	284.2	1.25	20	ac-sp^2
3	285.2	1.25	20	ac-sp^3
4	286.5	1.10	20	C-O

The peak numbers appear in Fig. 6.3

respectively [18, 19]. The proportion between both types of carbon (sp^2/sp^3) is found to be 3.45 ± 0.15 through the profile.

From the XPS analysis performed, we conclude that FIBID-W deposits are formed by metallic W filaments whose surface has bonding to carbon atoms, forming WC.

6.5 Superconducting Electrical Properties of Micro- and Nanowires

The aim of these experiments is first to characterize "bulk-type" W deposits in the micrometer range in order to investigate if narrow wires in the nanometer range continue to keep the same performance as the bulk ones. The best resolution obtained corresponds to width $\sim$ 100 nm, determined by SEM inspection (see Fig. 6.4).

In Table 6.4, some relevant data of the wires studied in the present work are summarized. Four microwires (μWs: M_1–M_4) and four nanowires (NWs: N_1–N_4) have been investigated.

6.5.1 Critical Temperature of the Wires

In Fig. 6.5a, the temperature dependence of the resistivity of the M_3 μW (grown at 20 kV/0.76 nA) is shown. It is found that $T_C = 5.4$ K, where the critical temperature

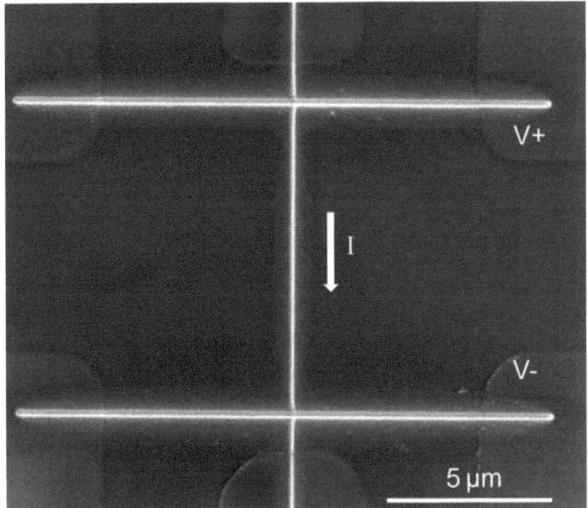

Fig. 6.4 SEM image of one deposited NW (*vertical line*). The additional two horizontal lines are made to perform four wire electrical measurements

Table 6.4 Data of the growth conditions (V_{BEAM}, I_{BEAM}), dimensions (width, section area), and physical properties (T_C and room-temperature resistivity) of all the wires investigated

Sample	V_{BEAM} (kV)	I_{BEAM} (nA)	Width (µm)	Section area (µm^2)	T_C (K)	ρ_{300K} (µΩ cm)
M_1	5	1	2	1.97	5.3	163
M_2	10	1.1	2	1.62	5.4	147
M_3	20	0.76	2	2.10	4.7	209
M_4	30	1	2	1.64	4.9	229
N_1	30	0.01	0.24	0.036	5.1	309
N_2	30	0.01	0.17	0.017	5.0	295
N_3	30	0.01	0.11	0.016	4.8	533
N_4	30	0.01	0.15	0.015	4.8	633

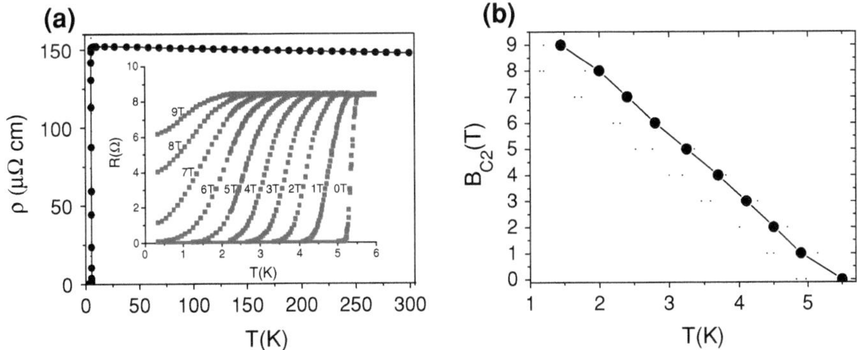

Fig. 6.5 For the M_3 microwire, **a** resistivity as a function of temperature. The *inset* shows measurements of the resistivity under various applied magnetic fields. **b** The critical field, B_{C2}, as a function of temperature

is defined at 90% of the transition ($\rho/\rho_n = 0.9$). From the measurements of the resistivity under several magnetic fields, the critical field, B_{C2}, has been estimated and shown in Fig. 6.5b, being of the order of a few Tesla, with a typical temperature dependence. These results coincide perfectly with those by Sadki et al. [1]. For all the μWs (grown at different conditions as shown in Table 6.4.), the measured T_C ranges between 4.8 and 5.4 K and seems to correlate with the room-temperature resistivity as proposed by Li et al. [6].

In Fig. 6.6, the resistance-versus-temperature behavior of one NW (N_2) is shown. The resistance starts to decrease at 5.2 K and becomes null at 4.8 K. The transition temperature (≈ 5 K) and transition width (≈ 0.4 K) are similar to those of μWs, indicating good sample stoichiometry and homogeneity. It is remarkable that a high T_C is maintained in the NWs, which opens the route for the use of this superconducting material in nanodevices working at temperatures above liquid Helium.

The values of T_C and room temperature resistivity for all the wires (micro and nano) are indicated in Fig. 6.7. It is generally observed that the NWs show slightly

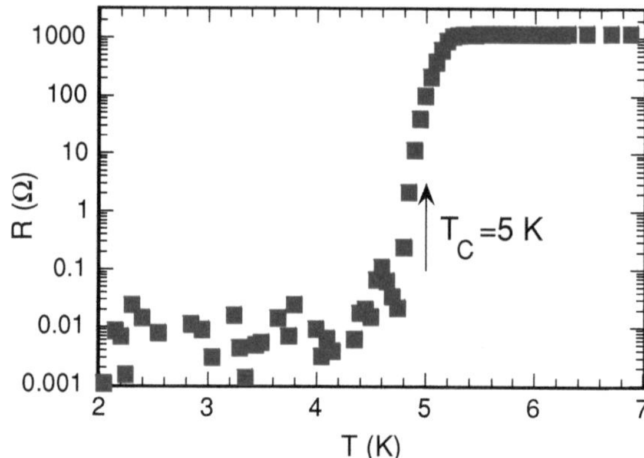

Fig. 6.6 Resistance (in log scale) versus temperature of N_2 (width = 170 nm), indicating $T_C = 5 \pm 0.2$ K

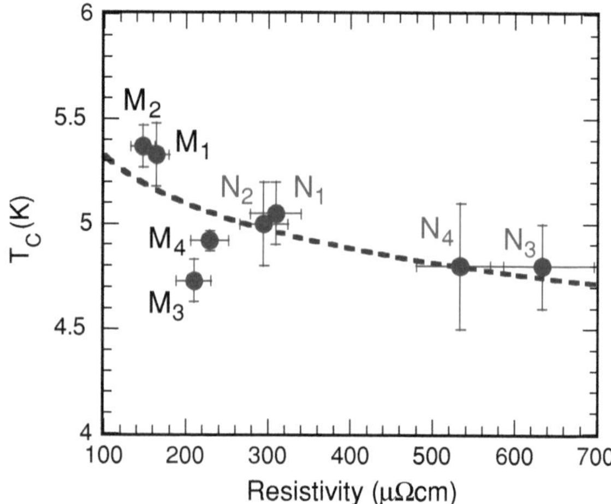

Fig. 6.7 T_C versus resistivity of all the microwires (M_1, M_2, M_3, M_4) and nanowires (N_1, N_2, N_3, N_4) studied in the present work. Whereas the nanowires show higher resistivity, T_C ranges in all cases between 4.7 and 5.4 K, with a slight tendency to decrease as a function of the resistivity

higher resistivity than the μWs, but without evident influence on the T_C value. More work is needed to understand the correlation between the room-temperature resistivity and T_C.

6.5.2 Critical Field of the Nanowires

The resistance in the NWs has been studied at fixed temperature as a function of the applied magnetic field with the aim of determining the range of magnetic fields where the NWs can be used for superconducting applications. As an example of the results obtained, we focus on the behavior of the N_2 sample at $T = 2$ K, represented in Fig. 6.8a. The resistance is null below B $\sim$2 T and increases continuously until reaching the normal-state resistance. The value of B_{C2} is found to be $\sim$8 T, which is the same as found in the μWs, as shown in the inset of Fig. 6.5a. All the studied NWs show similar behavior to this, as can be observed in Fig. 6.8b, where NWs N_1 to N_3 are compared normalizing the resistance to the normal-state value. These results demonstrate the good translation of the SC properties of FIBID-W deposits to the nanometric scale. A number of interesting features appearing in Fig. 6.8 will be discussed and systematically analyzed in the future. Here, we simply highlight the onset of finite resistance at relatively low magnetic field, the non-linear resistance before reaching the normal-state resistance, and the dependence on the NW thickness of the starting magnetic field with non-zero resistance.

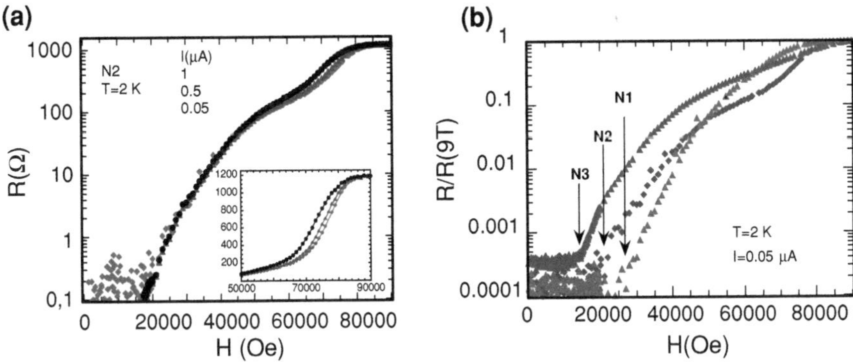

Fig. 6.8   a Resistance (in log scale) versus applied magnetic field at $T = 2$ K of N_2. Three different currents have been used (1, 0.5, 0.05 μA) and the *inset* shows in linear scale the different behavior produced by the increasing current. **b** At $T = 2$ K, resistance normalized to the value at $H = 9$ T as a function of the magnetic field in nanowires N_1, N_2, and N_3 using 0.05 μA current in all cases

6.5.3 Critical Current of the Nanowires

Another important issue to be studied in these NWs, bearing in mind certain applications, is the value and dependence of the critical current. The results obtained on the N_2 sample are shown in Fig. 6.9a to illustrate the dependence found. At T = 2 K, the critical current in this NW is of the order of 20 μA, decreasing for higher temperatures. More focused studies would be required to investigate if these nanodeposits show the "peak effect" of the critical current observed in other superconductors [20, 21]. In Fig. 6.9b we represent the resistance at T = 2 K normalized to the normal state as a function of the current density in several nanowires. At that temperature, the value of critical current density ranges between 0.06 and 0.09 MA/cm^2. It could be deduced that the dispersion in the measured values are to be ascribed to slight differences in composition, the amount of defects, error in the determination of the sectional area, etc. Moreover, these results suggest that the critical current density decreases as the NW width decreases (from 240 nm for N_1 down to 170 nm for N_2 and 110 nm for N_3). This point should be addressed in more detail in future.

6.6 Study of FIBID-W by Scanning-Tunneling-Spectroscopy

In this section we will show a brief summary of the results obtained as a consequence of the collaboration established between our group and Prof. Sebastián Vieira, Dr. Hermann Suderow and Dr. Isabel Guillamón, at the Universidad Autónoma de Madrid [this work is part of the thesis of Isabel Guillamón 22].

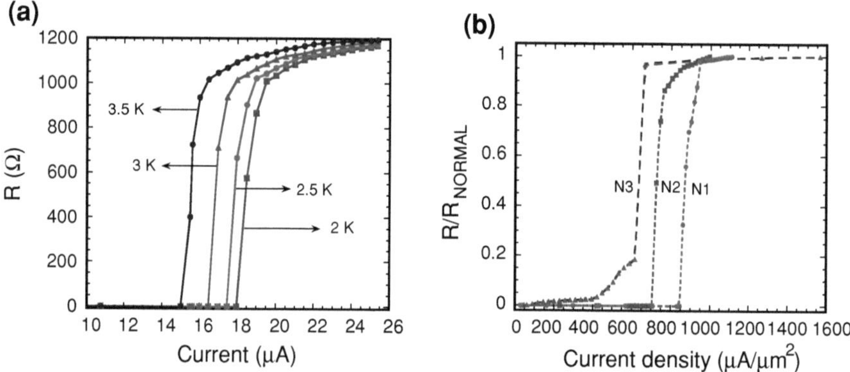

Fig. 6.9 a For N_2 sample, resistance versus applied current at different temperatures. As expected, the critical current decreases as T_C is approached. **b** For several NWs, dependence of the resistance (normalized to the value in the normal state) with the injected current density at T = 2 K

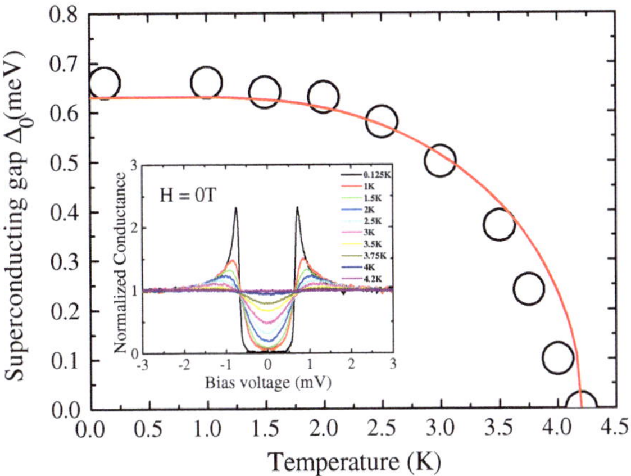

Fig. 6.10 Temperature dependence of the superconducting gap obtained from fits to the tunneling spectra shown in the *inset*. The *red line* is the BCS expression, taking $\Delta_0 = 1.76\ kT_c$ and $T_c = 4.15$ K

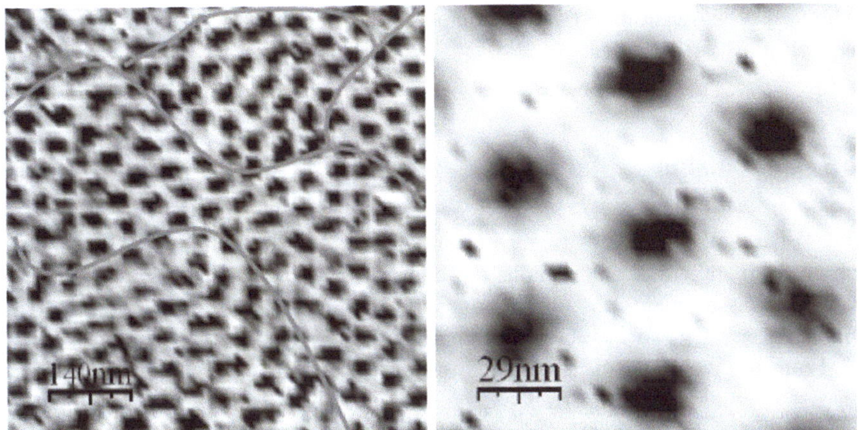

Fig. 6.11 Vortices lattice at 1 T, obtained from STS images constructed from the Fermi level conductance (shown as raw images, without any filtering or smoothing effects). Left panel shows a large STS image, and right panel shows a single hexagon. Small linear surface depressions found in the topographic images are marked as *lines*. The STS images of the right panels are made far from surface depressions, and the corresponding topographic STM images are absolutely flat and featureless, with surface roughness below about 0.5 nm

A FIBID-W micrometric sample (30 μm × 30 μm), deposited on a Au–Pd substrate, was studied by Scanning-Tunneling- Microscopy (STM) and Spectroscopy (STS). STS allows spectroscopy studies with a spatial resolution down to the atomic scale [23], allowing a direct mapping of the vortex lattice [24].

The first outstanding result was that tunneling experiments could be done in a FIBID-W sample that had been exposed to ambient conditions for 1 day (normally extremely clean surfaces are necessary for this kind of experiments), presenting a very homogeneous surface. By a detailed set of measurements, it was found that the density of states of this superconductor could be perfectly fitted to the Bardeen–Cooper–Schrieffer (BCS) theory for superconductors (see Fig. 6.10).

Until experiments were performed in this material, atomic level STM/S experiments over large surfaces in a SC were still lacking. This work shows that the amorphous FIBID-W system is the first reference s-wave BCS system, "the long sought paradigm for STM/S studies".

Furthermore, under an applied magnetic field, a well defined Abrikosov lattice was observed, being systematically studied as a function of temperature and field, by STS. In Fig. 6.11, an example of the vortex lattice is shown, with the vortices appearing bunched close to some linear surface depressions observed in the topography. By direct STS imaging, it was possible to observe the melting of the 2D-lattice into an isotropic liquid by means of two intermediate Kosterlitz–Thouless type transitions. The smectic phase had never been observed before in an equilibrium vortex lattice.

6.7 Conclusions

The FIBID-W nanodeposits were characterized by HRTEM. The amorphous character of this material explains the high T_c.

XPS measurements show a homogeneous composition, formed by metallic W filaments, with its surface bonded to carbon atoms, forming WC.

From the electrical measurements in microwires and nanowires, we conclude that superconducting properties are retained by the narrowest wires, opening up applications at the nanoscale. In particular, the values found for T_C, for the critical field H_{C2}, and for the critical current make these narrow wires promising for nanometric superconducting-based and hybrid devices.

Finally, the STS study done by the group of Prof. Sebastián Viera, shows outstanding properties for this material. The FIBID-W nanodeposit is the first fully isotropic s-wave superconductor that can be studied in detail down to atomic level, allowing the direct observation of the melting of the 2D-vortex lattice.

From these conclusions, it can be anticipated that FIBID-W could be feasibly used to build SC electronic devices, due to its unique characteristics and the controllable and flexible growth at the nanometric scale. For example, as logic elements in quantum computing, for the fabrication of micro/nano-SQUIDs at targeted places, for the fabrication of resistance-free nanocontacts, or for building hybrid FM-SC nanocontacts for the study of the spin polarization of the FM. The first steps in these directions are being taken.

References

1. E.S. Sadki, S. Ooi, K. Hirata, Focused-ion-beam-induced deposition of superconducting nanowires. Appl. Phys. Lett. **85**, 6206 (2004)
2. E.S. Sadki, S. Ooi, K. Hirata, Focused-ion-beam-induced deposition of superconducting thin films. Physica C. **426–431**, 1547 (2005)
3. J.W. Gibson, R.A. Hein, Superconductivity of tungsten. Phys. Rev. Lett. **12**, 688 (1964)
4. I.J. Luxmoore, I.M. Ross, A.G. Cullis, P.W. Fry, J. Orr, P.D. Buckle, J.H. Jefferson, Low temperature electrical characterization of tungsten nano-wires fabricated by electron and ion beam induced chemical vapor deposition. Thin Solid Films **515**, 6791 (2007)
5. D. Spoddig, K. Schindler, P. Rodiger, J. Barzola-Quiquia, K. Fritsch, H. Mulders, P. Esquinazi, Transport properties and growth parameters of PdC and WC nanowires prepared in a dual-beam microscope. Nanotechnology **18**, 495202 (2007)
6. W. Li, J.C. Fenton, Y. Wang, D.W. McComb, P.A. Warburton, Tunability of the superconductivity of tungsten films grown by focused-ion-beam direct writing. J. Appl. Phys. **104**, 093913 (2008)
7. M.S. Osofsky, R.J. Soulen Jr., J.H. Claassen, G. Trotter, H. Kim, J.S. Horwitz, New insight into enhanced superconductivity in metals near the metal-insulator transition. Phys. Rev. Lett. **87**, 197004 (2001)
8. A.Y. Kasumov, K. Tsukagoshi, M. Kawamura, T. Kobayashi, Y. Aoyagi, K. Senba, T. Kodama, H. Nishikawa, I. Ikemoto, K. Kikuchi, V.T. Volkov, Yu.A. Kasumov, R. Deblock, S. Guéron, H. Bouchiat, Proximity effect in a superconductor-metallofullerene-superconductor molecular junction. Phys. Rev. B **72**, 033414 (2005)
9. A. Shailos, W. Nativel, A. Kasumov, C. Collet, M. Ferrier, S. Guéron, R. Deblock, H. Bouchiat, Proximity effect and multiple Andreev reflections in few-layer graphene. Europhys. Lett. **79**, 57008 (2007)
10. W.L. Bond, A.S. Cooper, K. Andres, G.W. Hull, T.H. Geballe, B.T. Matthias, Superconductivity in films of β tungsten and other transition metals. Phys. Rev. Lett. **15**, 260 (1965)
11. R.H. Willens, E. Buehler, The superconductivity of the monocarbides of tungsten and molybdenum. Appl. Phys. Lett. **7**, 25 (1965)
12. M.M. Collver, R.H. Hammond, Superconductivity in "amorphous" transition-metal alloy films. Phys. Rev. Lett. **30**, 92 (1973)
13. S. Kondo, Superconducting characteristics and the thermal stability of tungsten-based amorphous thin films. J. Mater. Res. **7**, 853 (1992)
14. H. Miki, T. Takeno, T. Takagi, A. Bozhko, M. Shupegin, H. Onodera, T. Komiyama, T. Aoyama, Superconductivity in W-containing diamond-like nanocomposite films. Diam. Relat. Mater. **15**, 1898 (2006)
15. R. Córdoba, Thesis, to be published in 2011 at the University of Zaragoza, rocorcas@unizar.es
16. J. Luthin, Ch. Linsmeier, Carbon films and carbide formation on tungsten. Surf. Sci. **454–456**, 78 (2000)
17. M. Xu, W. Zhang, Z. Wu, S. Pu, Evolution mechanism of nanocrystalline tungsten-carbon and effects on tungsten implanted amorphous hydrogenated carbon. J. Appl. Phys. **102**, 113517 (2007)
18. J. Díaz, G. Paolicelli, S. Ferrer, F. Comin, Separation of the sp^3 and sp^2 components in the C1s photoemission spectra of amorphous carbon films. Phys. Rev. B **54**, 8064 (1996)
19. R. Haerle, E. Riedo, A. Pasquarello, A. Baldereschi, sp^2/sp^3 Hybridization ratio in amorphous carbon from C 1s core-level shifts: X-ray photoelectron spectroscopy and first-principles calculation. Phys. Rev. B **65**, 045101 (2002)
20. M.C. Hellerqvist, D. Ephron, W.R. White, M.R. Beasley, A. Kapitulnik, Vortex dynamics in two-dimensional amorphous $Mo_{77}Ge_{23}$ films. Phys. Rev. Lett. **76**, 4022 (1996)

21. N. Kokubo, T. Asada, K. Kadowaki, K. Takita, T.G. Sorop, P.H. Kes, Dynamic ordering of driven vortex matter in the peak effect regime of amorphous MoGe films and 2H-NbSe2 crystals. Phys. Rev. Lett. **75**, 184512 (2007)
22. I. Guillamón, Thesis, published in 2009 at the Universidad Autónoma de Madrid, isabel. guillamon@uam.es
23. Ø. Fischer, M. Kugler, I. Maggio-Apprile, C. Berthod, Scanning tunneling spectroscopy of high-temperature superconductors. Rev. Mod. Phys. **79**, 353 (2007)
24. H.F. Hess, R.B. Robinson, R.C. Dynes, J.M. Valles, J.V. Waszczak, Scanning-tunneling-microscope observation of the Abrikosov flux lattice and the density of states near and inside a fluxoid. Phys. Rev. Lett. **62**, 214 (1989)

Chapter 7
Magnetic Cobalt Nanowires Created by FEBID

This chapter includes the characterization of cobalt nanostructures created by FEBID. Microstructural and spectroscopic measurements show that the deposits are extremely pure (> 90%), when relatively high beam currents are used, in sharp contrast with usual concentrations obtained using this technique (below 50%), which is ascribed to a heating effect. A wide range of magnetotransport measurements have been performed, evidencing the high Co percentage in these deposits. By using spatially-resolved Magneto-Optical-Kerr magnetometry, the reversal of the magnetization has been systematically studied in single wires with different aspect ratios. Moreover, nanowires exhibiting good domain wall conduit behavior (propagation field much lower than nucleation field) have been created.

7.1 Previous Results for Local Deposition of Magnetic Materials Using Focused Beams

The capability of locally growing magnetic devices with high resolution is a longed-for result, since its realization would have a high impact in fields such as magnetic storage, or magnetic sensing, among others.

The possibility to create magnetic nanometric-sized structures using a focused technique (FIBID or FEBID) seems thus very attractive for this purpose, since these techniques posses a high flexibility and resolution, having a perfect compatibility with other lithography techniques. These are obvious advantages with respect to multi-step processes involving growth and subsequent patterning. However, these techniques have not been for the moment exploited as much as, for example, EBL or direct patterning by FIB for the growth of functional nanometric elements. The main reason for it is the low purity of the deposits (rarely exceeding the 50% atomic), as a consequence of the absence of the complete low dissociation rate of the gas precursor molecules, especially for FEBID (Sect. 1.5.3). Organic

A. Fernandez-Pacheco, *Studies of Nanoconstrictions, Nanowires and Fe₃O₄ Thin Films*, Springer Theses, DOI: 10.1007/978-3-642-15801-8_7,
© Springer-Verlag Berlin Heidelberg 2011

precursors are usually employed, resulting in a big amount of carbon mixed with the metallic element. Thus, properties deviate radically from those ones desired (examples of how materials deposited by these techniques deviate from the metallic behavior can be found in Chaps. 5 and 6 of this thesis, for Pt- and W-based deposits, respectively).

Some results can be found in literature concerning the deposition of magnetic nanostructures by these techniques. For iron, several gas precursors have been used: $Fe(CO)_5$ [1, 2, 3, 4], $Fe_3(CO)_{12}$ and $Fe_2(CO)_9$ (Ref. [5]). For cobalt, the same gas precursor is mainly used: $Co_2(CO)_8$ [6–11] (except Ref. [9], all the works cited were done with FEBID technique).

Due to the big fraction of carbon normally present in the deposits, it is crucial to explore the functionality of these nanomaterials, by means of a thorough magnetic characterization. Before this thesis, studies of the magnetism of single nanodeposits had been carried out, either in an indirect way, by electrical magneto-transport measurements [7, 11], or by local probes, such as Magnetic Force Microscopy [6, 10, 11] or electron holography [1, 2, 3].

7.2 Experimental Details

The FEBID magnetic nanodeposits were fabricated at room temperature in the dual-beam equipment detailed in Sect. 2.1.2, which uses a 5–30 kV FEG-Column. The $Co_2(CO)_8$ precursor gas is brought onto the substrate surface by a GIS, where it becomes decomposed by the FEB. Common parameters for this deposition process are: Precursor gas temperature $=$ Room Temperature, Vol/dose $= 5 \times 10^{-4}$ $\mu m^3/nC$, dwell time $= 1$ μs, beam overlap $= 50\%$, base chamber pressure $= 10^{-7}$ mbar, process pressure $= 3 \times 10^{-6}$ mbar, vertical (horizontal) distance between GIS needle and substrate $= 135$ (50) μm.

7.2.1 Compositional Analysis by EDX

EDX measurements we performed in the setup detailed in Sect. 2.3.1, for deposits created with beam energy and current in the ranges of 5–30 kV and 0.13–9.5 nA, respectively. The selected electron beam energy for the microanalysis is 20 kV.

7.2.2 HRTEM

The high-resolution transmission electron microscopy (HRTEM) study was carried out using a FEG-TEM Jeol 1010 equipment® operated at 200 kV (point to point resolution 0.19 nm) on 50 nm-thick Co nanodeposits directly grown on TEM grids for electron transparency in the experiment.

7.2.3 Electrical Measurements of Wires

Rectangular wires were deposited on previously-micropatterned $SiO_2//Si$, where Al or Ti pads were evaporated after an optical lithography process, to realize contact pads for the magnetotransport measurements (see details of lithography in Sect. 2.1.1).

These measurements were done in two different equipments, depending on the geometry of the applied magnetic field with respect to the thin film plane. For fields perpendicular to the thin film plane, a commercial PPMS from Quantum Design® was used, with fields up to 9 T and in a temperature range 2 K < T < 300 K. In this setup, perpendicular magnetoresistance and Hall effect were measured. For fields in the plane, the measurements were done in the CCR explained in Sect. 2.2.1. In this case, different geometries between the current and the magnetic field were established.

7.2.4 Spatially-Resolved MOKE Measurements

The MOKE measurements were performed at the Imperial College (London), in the group of Professor Russell Cowburn, under the supervision of Dr. Dorothée Petit. This setup (details in Sect. 2.4) measures the longitudinal-Kerr effect, with a laser probe of ~ 5 μm, allowing to measure hysteresis loops measurements of single nano-objects, with for instance, sensitivities of around 10^{-12} emu in the case of permalloy [12].

7.2.5 AFM Measurements

The profiles of some nanostructures were measured using contact mode atomic force microscopy.

7.3 Compositional (EDX) and Microstructural (HRTEM) Characterization

We have performed in situ EDX microanalysis of several Co nanodeposits grown under different combinations of electron-beam energy and current, in order to study any correlation between the growth conditions and the nanodeposit composition. A summary of the results is displayed in Table 7.1. The set of results clearly demonstrates that the main parameter governing the nanodeposit content is the beam current. For beam currents above ~ 2 nA, the Co content in atomic

Table 7.1 EDX results of FEBID-Co nanodeposits grown under different conditions

E_{BEAM} (keV)	I_{BEAM} (nA)	% C (atomic)	% O (atomic)	% Co (atomic)
5	0.4	15.1 ± 0.3	2.0 ± 0.2	82.9 ± 0.5
	1.6	13.1 ± 1.7	3.7 ± 0.7	83.3 ± 1.0
	6.3	7.2 ± 1.0	1.4 ± 0.4	91.4 ± 0.6
10	0.13	8.9 ± 0.2	10.2 ± 0.2	80.9 ± 0.2
	0.54	5.3 ± 1.1	2.7 ± 0.7	91.9 ± 1.8
	2.1	3.9 ± 0.9	1.4 ± 0.7	94.8 ± 1.6
	8.4	5.2 ± 0.1	1.2 ± 0.2	93.6 ± 0.3
18	9.2	9.3 ± 0.6	2.4 ± 0.3	88.3 ± 0.7
30	0.15	19.2 ± 0.5	19.3 ± 0.2	61.5 ± 0.5
	9.5	3.2 ± 0.2	0.2 ± 0.1	96.6 ± 0.4

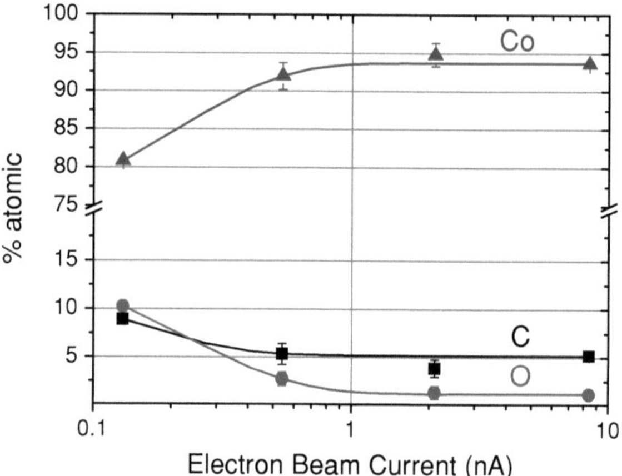

Fig. 7.1 Atomic percentage of Co (*triangles*), C (*squares*), and O (*dots*) as a function of the electron-beam current in nanodeposits grown with the FEBID technique at beam energy of 10 kV using $Co_2(CO)_8$ as the precursor gas. Details are given in the text. *Lines* are guides to the eye

percentage is around 90% or more, whereas for lower currents the Co content is typically around 80% or less. The rest of constituents of the nanodeposits are C and O.

As an example, in Fig. 7.1, the Co, C, and O content is represented as a function of the beam current in deposits performed at 10 kV beam energy. The Co content increases monotonously from about 80% for a beam current of 0.13 nA to about 95% for beam currents above 2 nA.

As far as we know, Co percentages above 90% have not been reported so far [6–8, 10, 11]. A previous study by Utke et al. [8] also found an increase in the Co content as a function of the beam current, as it is our case. They reported a maximum Co concentration around 80%, but currents above 80 nA were necessary for this high concentration. For currents in the range of our study, Co percentages were always found to be below 60% (we must notice in this point that the beam current is a crucial parameter in the electron beam resolution). It has been argued that beam-induced heating is the main effect responsible for the increase of metallic content with current [8, 13]. The precursor gas molecule, $Co_2(CO)_8$, can

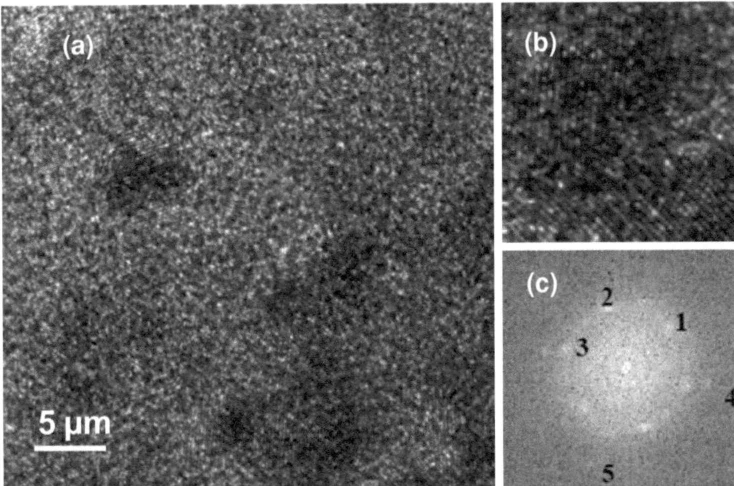

Fig. 7.2 a HRTEM images of one FEBID-Co deposit grown on a TEM grid, under 5 kV and 98 pA conditions. **b** Detail of the image, where crystallographic planes are well observed. **c** FFT of image (**b**), where crystallographic planes [1: (10–10) HCP, 2: (111) FCC, 3: (101) FCC, (10–12) HCP, (11–20) HCP] are observed

be thermally decomposed at relatively low temperature, 60°C [14]. As a consequence, the use of high current will favor local heating, enhancing the precursor molecule dissociation and the high Co content. The electron beam source used is a field-emission gun, which will likely produce smaller beam spots on the substrate than in the case of thermionic-tungsten-filament sources, employed for example in Ref. [8]. Thus, the electron beam current will be focused on a smaller area, producing larger local heating effects. We put forward that this is the reason for the higher Co content in our nanodeposits compared to previous studies. Recent results in our group by R. Córdoba (Micr. Eng, 2010) using a micro-heater during deposition confirm this hypothesis.

HRTEM measurements (this piece of work forms part of the thesis of Rosa Córdoba (University of Zaragoza: rocorcas@unizar.es)) indicate a high purity in the deposits. In Fig. 7.2 the microstructure of one deposit, grown at 5 kV/98 pA, is shown. FEBID-Co is formed by small cobalt crystalline grains, of size $\sim$2–5 nm. The grains are randomly oriented, with HCP and FCC structures, as electron-diffraction patterns indicate.

7.4 Magnetotransport Measurements of FEBID-Co Nanowires

In this section, we will demonstrate that a high content of Co is required to obtain high-quality magnetotransport properties. For this study, the Co-based deposits

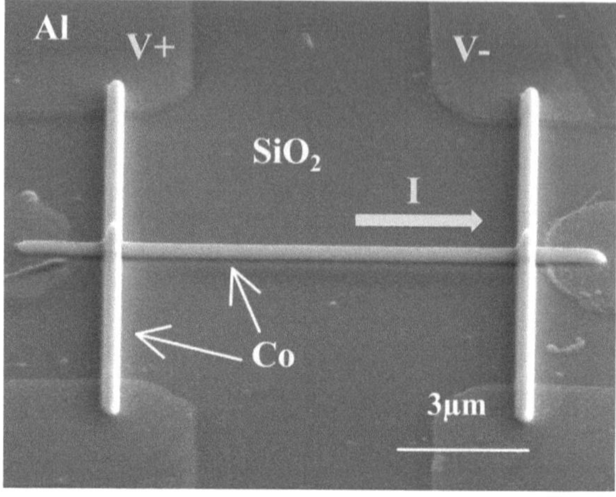

Fig. 7.3 SEM image of one of the created devices. The current flows through the horizontal deposit, whereas transversal deposits are performed for four-probe measurements. Deposits are done on top of a SiO$_2$ substrate in which Al pads were previously patterned by conventional optical lithography techniques

were designed to bridge a 12 µm gap (horizontal line in Fig. 7.3). Typical dimensions for the deposits are length = 9 µm × width = 500 nm × thickness = 200 nm. Additional transversal deposits were designed to determine, by conventional 4-probe measurements, the longitudinal and transversal components of the resistivity tensor.

7.4.1 Magnetotransport Properties of Cobalt NWs Grown at High Currents

As an example, we will show hereafter the results obtained on a NW grown with a beam energy of 10 kV and a beam current of 2.1 nA, thus having a high Co content (around 95%). A wide range of magnetic field-dependent transport measurements have been performed. The relevance of the surface oxidation of these cobalt NWs for the magnetotransport properties has been evaluated by preparing some samples with a 5–10 nm FEBID Pt capping layer. This small FEBID-Pt thickness together with its high resistivity (see Sect. 5.2), guarantees a high resistance in parallel with the Co lines, assuring no perturbation in the Co conduction, as it was subsequently checked. After several days in air, no changes were found in the resistance of any NW (either if they were covered or not). We expect that only the usual passivation layer (~ 1–2 nm) at the topmost surface will be present in the non-covered NWs grown at high beam currents, without any significant influence on the obtained results. This is not the general case in other deposits created by focused beams (see, for instance, Sect. 5.1.3.2).

In Fig. 7.4 we can see a typical resistivity-versus-temperature measurement for this NW. The behaviour is not, as it is the general rule in FEBID deposits, semiconducting-like (Sect. 5.2.2.3), but fully metallic, confirming that the residual amount of carbon in the sample is minimal. The resistivity at room temperature is

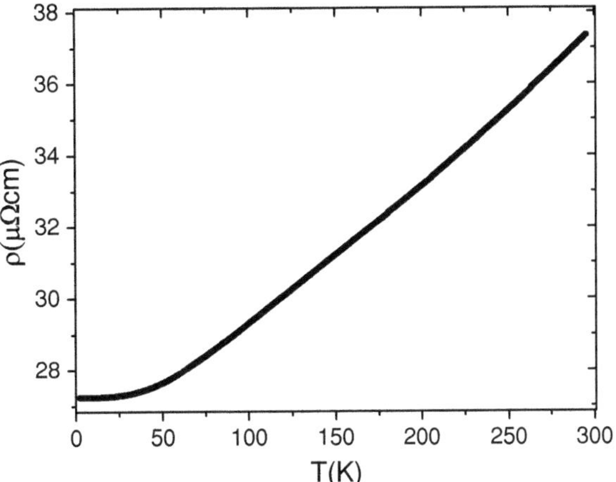

Fig. 7.4 Resistivity of a Co nanowire grown at 10 kV beam energy and high current, 2 nA, as a function of temperature. Fully metallic behavior is observed

of the order of 40 $\mu\Omega$ cm, only a factor of 7 larger than in bulk polycrystalline cobalt. As will be discussed later, this value of resistivity is just related to the small grain size of the FEBID deposit. Another NW grown at 10 kV and 0.54 nA, with Co content around 92%, also displays similar resistivity value. These FEBID-Co deposits are the most conductive ones found in literature. The RRR (= ρ_{300K}/ρ_{2K}) is about 1.3, in agreement with results for polycrystalline cobalt thin films [15].

In Fig. 7.5 we show the Hall effect resistivity for the NW as a function of field for the maximum and minimum temperature measured. ρ_H increases at low fields until reaching magnetic saturation. For higher fields, only the ordinary part variation (ρ_{OH}) is observed, with contrary sign to the anomalous one (ρ_{AH}) (see Eq. 3.9). This behavior has also been observed in magnetron sputtered polycrystalline cobalt films [15]. The ordinary part is obtained by the standard method of fitting the slope at high fields, whereas the anomalous one is determined by extrapolating to H = 0 the fit of ρ_{OH} (see Fig. 7.5). From the different isotherms, we find that the relationship $\rho_{AH} \propto \rho^2$ holds in all cases, revealing an intrinsic origin of the anomalous Hall effect, as it is expected in this conductivity regime [16, 17]. From the intersection of the linear regions of the ordinary and the anomalous Hall effect [18] (1.42 T at 300 K) and taking into account the demagnetizing factor for the created geometry, we find a saturation magnetization $M_s = 1329 \pm 20$ emu/cm^3, corresponding to a 97% of the bulk Co value. We would like to remark that such high magnetization has never been reported before for Co nanodeposits created by FEBID.

From the ordinary part at 2 K, the density of electrons is calculated to be $N_e = 7 \times 10^{-28}$ m^{-3}. Using this value and the experimental value of the resistivity at low temperature, $\rho(0) \approx 27$ $\mu\Omega$ cm, one can estimate the mean-free-path as in reference 284:

$$l(0) = \frac{\hbar k_F}{n_e e^2 \rho(0)} \approx 4 \text{ nm} \tag{7.1}$$

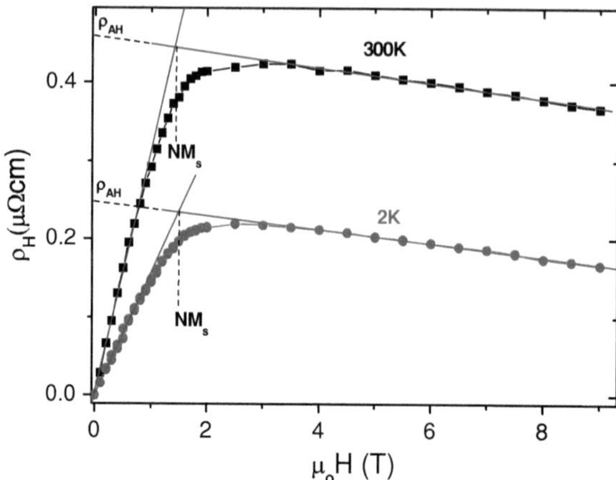

Fig. 7.5 Hall resistivity at 300 K and 2 K in the same Co nanowire shown in Fig. 2. The magnetic character is manifested by the presence of the anomalous Hall effect. The ordinary part is determined by the slope at high H, whereas the anomalous constant is calculated by extrapolating this slope at zero field. From the intersection of the linear regions of the ordinary and the anomalous Hall effect the value of NM_S (N demagnetising factor, M_s saturation magnetization) is obtained, which allows the determination of M_s (see text for details)

Fig. 7.6 Magnetoresistance measurements at 300 K in longitudinal, transversal and perpendicular geometries of the same Co NW as in Figs. 7.5 and 7.4

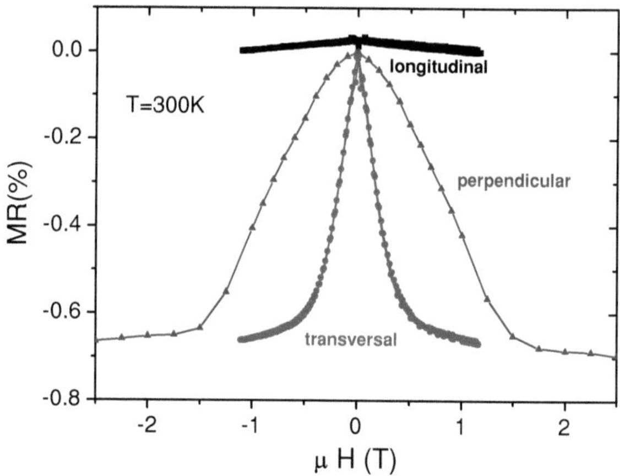

This value is of the order of the grain size determined by HRTEM images (see Sect. 7.3), suggesting that the scattering is mainly produced at the grain boundaries between crystalline grains.

In Fig. 7.6, MR measurements at 300 K are shown with three different geometries (shown in Fig. 3.9): H perpendicular to the thin film plane (PG), H in plane and parallel to the current I (LG), H in plane and perpendicular to I (TG). The measured MR in the three geometries is in perfect agreement with previous results in Co NWs fabricated by an EBL process [19, 20]. The magnetization prefers to align along the NW axis due to the strong shape anisotropy (length $\approx 18 \times$ width $\approx 45 \times$ thickness). The differences in the saturation fields for TG and PG are due to the width-thickness ratio (>1). The slight decrease of the MR at high magnetic fields is caused by small misalignments of the wire with H. The different magnetoresistance observed between the LG and any of the others is caused by the AMR (see Sect. 3.5. for details of AMR). The value at 300 K is positive, around 0.65–0.7%, as previously found in Co NWs [19, 20] and thin films [21]. Supposing coherent rotation when the field is applied in the PG, the anisotropy constant in this geometry, K_u^{PG}, can be estimated. From Ref. [19]:

$$R(H) = R_{\parallel} - \left(R_{\parallel} - R_{\perp}\right)\left(\frac{\mu_0 M_s H}{2K_u^{PG}}\right)^2 \tag{7.2}$$

where, $R_{\parallel}$ and $R_{\perp}$ are, respectively, the resistances at zero field and after magnetic saturation along the perpendicular direction. At 300 K, $K_u^{PG} = 7.9 \times 10^5$ J/m^3 is obtained. In the TG, however, the dependence is far from being quadratic, and the same analysis cannot be performed. This suggests that the mechanism towards saturation in this configuration is not just by coherent rotation of the magnetization. Further experiments with specific magnetic techniques are requested to investigate in detail the exact mechanisms of magnetization reversal in these NWs.

The AMR was further studied as a function of temperature by means of the PHE in a 45° configuration between I and H, where the transversal voltage is maximum (Sect. 3.6.1). In Fig. 7.7a we can see the transversal resistivity ρ_{xy} in this configuration for several isotherms. From ρ_{xy} and under saturation, the AMR ratios can be calculated using Eq. 3.8.

The AMR slightly increases when lowering temperature, from 0.65% at 300 K to 0.8% at 25 K, in good agreement with EBL-fabricated Co NWs [19, 20]. Figure 7.7b shows in detail ρ_{xy} in a low range of field in two selected isotherms. Abrupt switches occur around 150 Oe, the switching field increasing when temperature diminishes, as expected for an activation process [22].

7.4.2 Magnetotransport Properties of Cobalt NWs Grown at Low Beam Currents

For the sake of completeness, we will show the results obtained in one NW grown at 10 kV and low current, 0.13 nA, with an estimated cobalt content around 80%. As shown in Fig. 7.8, such a low cobalt content precludes the achievement of high-quality magnetotransport properties. The room temperature resistivity value, 10,800 $\mu\Omega$ cm, is more than 300 times larger than in the case of the NWs grown under high currents. Besides, the study of the resistivity as a function of temperature indicates a slightly semiconducting behavior, in sharp contrast to the full metallic behavior found for the NWs grown under high currents. The current-versus-voltage measurements performed at 25 and 300 K are linear, indicating an ohmic behavior. The MR measurements shown in the top inset of Fig. 7.8 indicate low MR ratios in

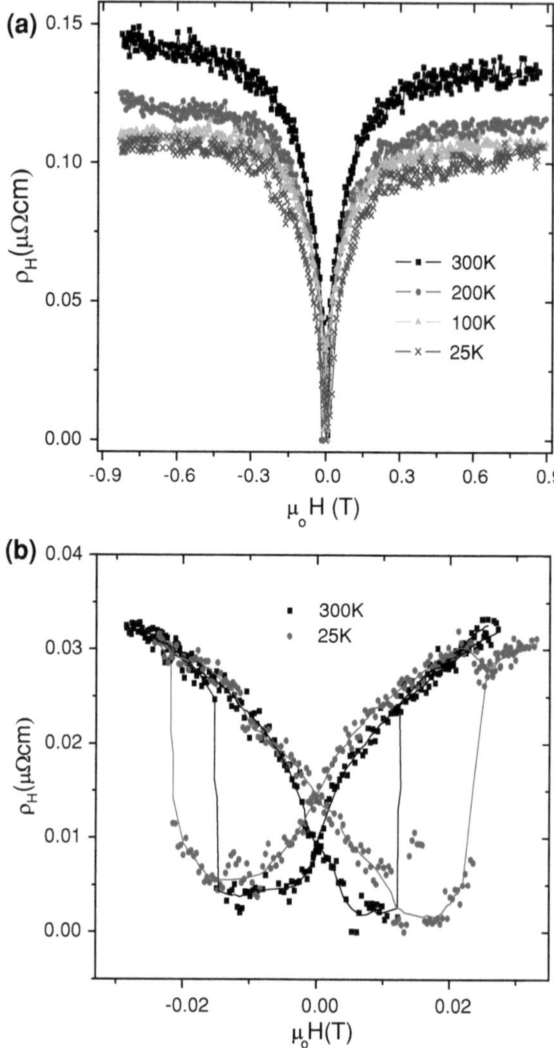

Fig. 7.7 a Planar Hall effect resistivity isotherms performed in the Co NW shown in Fig. 7.3. The magnetic field is applied in the substrate plane, forming 45° with the current, where the PHE voltage is maximum. From these measurements, the AMR has been determined as a function of temperature (see text for details). **b** Detail of (**a**) at low magnetic fields, showing abrupt magnetic switches as a consequence of the strong shape anisotropy and the magnetic field configuration. The *lines* are guides to the eye

the perpendicular and transversal configurations, about -0.08%, one order of magnitude smaller than those found for the NWs grown under high currents and those expected in bulk polycrystalline Co. Summarizing, nanodeposits with high Co content are required to obtain high-quality magnetotransport properties.

7.5 Systematic Study of Rectangular Wires

After having demonstrated that FEBID-Co is really pure, in this section, we will show a systematic study by means of spatially-resolved MOKE of the magnetic

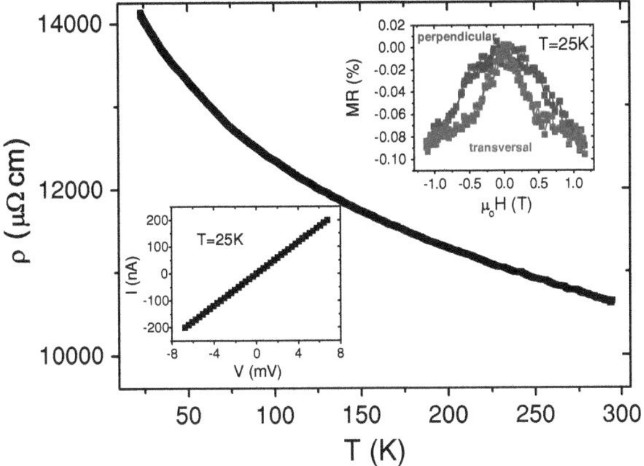

Fig. 7.8 Resistivity as a function of temperature of a Co NW grown at 10 kV beam energy and low current, 0.13 nA. The bottom inset shows the current-versus-voltage measurement at 25 K and the top inset displays magnetoresistance measurements at the same temperature. The high value of the resistivity, the semiconducting temperature dependence and the low MR values illustrate the degraded magnetotransport properties of NWs grown at low current

hysteresis loops of single wires in a wide range of aspect ratios, and grown under different beam currents. The dependence of the switching fields with the dimensions is studied, and compared with analogous structures fabricated by other techniques. This study tries to understand the mechanisms for the reversal of the magnetization for these structures. This has been under intense study for the last decade in nanoscale structures [19, 23–28] because the performance of devices based on these nanoelements depends on such magnetic switching.

7.5.1 Types of Structures Grown. Maximum Resolution Obtained

The deposition parameters were the following: precursor temperature = 23°C, beam voltage (V_{FEB}) = 5 kV, beam currents (I_{BEAM}) = 24, 6.3, 1.6 and 0.4 nA, substrate temperature = 22°C, dwell time = 1 μs, base chamber pressure $\approx 10^{-6}$ mbar, process pressure $\approx 3 \times 10^{-6}$ mbar, beam overlap = 50%, out of plane (in plane) distance between GIS needle and focus point on the substrate = 150 (50) μm. Two types of substrates were used for these experiments: a silicon wafer (type I), and a thermally oxidized Si substrate (200 nm-thick SiO_2 layer: type II). The dimensions of the rectangular lines are: lengths (L) = 8 or 4 μm (no significant difference was found for similar aspect ratios), and thicknesses (t) = 200 or 50 nm. The nominal widths (w) were chosen in order to have structures with aspect ratios (AR = L/w) varying from 1 up to 20. The study was completed by depositing single pixel lines using I_{BEAM} = 0.4 nA, in order to

Fig. 7.9 SEM images of
rectangular FEBID-Co
deposits with different aspect
ratios. A micron wide
parasitic halo deposit is
present around the main wire

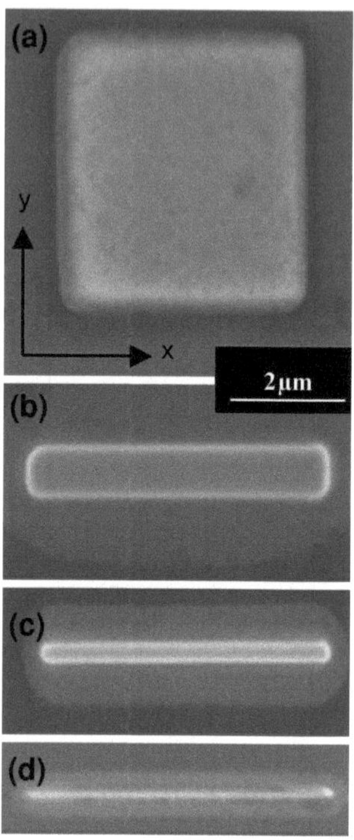

determine the best resolution achievable under the chosen conditions. The narrowest line was found to have a full width at half maximum (FWHM) of around 150 nm for a 40 nm-thick deposit (corresponding to an AR of 26 for L = 4 µm). Because of the interaction volume of the electrons with the substrate, this is larger than the FEB spot size [13, 29]. Some SEM images of lines deposited on top of both substrates are shown in Fig. 7.9. A parasitic deposit is visible around the structure (the extent of which, as determined by SEM images, ranges between 400 nm and 1 µm, slightly larger in the case of the type II-substrate). This halo is caused by BSE and their SE [7, 13, 29], a phenomenon similar to the one observed in EBL. A previous work reports a lower metal concentration in the halo when using this gas precursor, about 5 times smaller than in the main deposit [11].

7.5.2 AFM Investigation of the NWs Topography

The AFM profiles of some lines are shown in Fig. 7.10, together with a 3-dimensional image of the corresponding wire. The structures are highly uniform in width, whereas deviations up to 20% are observed in the thickness of the largest structures. The lateral shape of deposits grown under the same conditions are equivalent, but differences in the process pressure and beam focus conditions cause

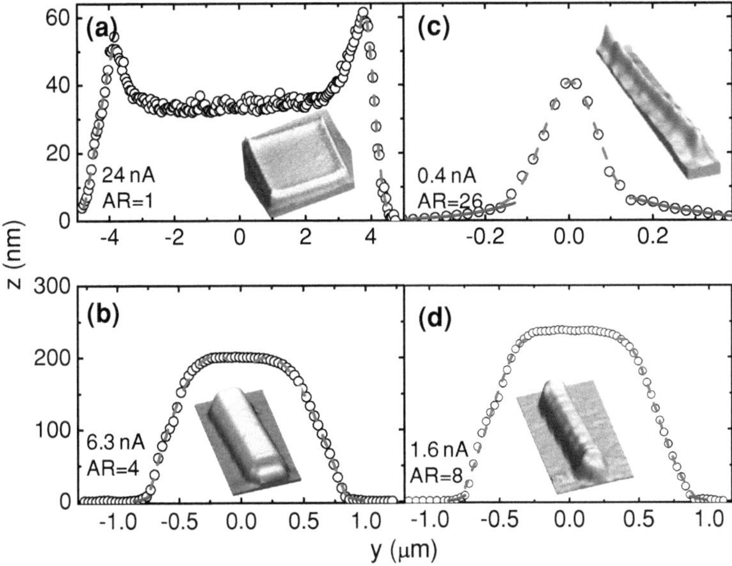

Fig. 7.10 Cross sectional profiles along the short axis of the wires, extracted from the AFM images (*insets*). The beam currents used as well as the AR (= L/w) of the wires are indicated. In (**a**) the profile is associated to the DER, whereas the rest of the profiles can be explained by a PLR. Experimental points (*open circles*), as well as the fits of the different part of the profiles to $z_2(y)$ (*dashed line*) and $z_3(y)$ (*solid line*), are shown

changes in the thickness of up to 30%. The profiles are highly dependent on the beam spot used for fabrication [13, 29]. In Fig. 7.10a, a 8 μm × 8 μm × 50 nm structure, grown with a beam current I_{BEAM} of 24 nA, shows a non-uniform height distribution, with edges higher than the centre. This effect has been previously observed, for instance, when depositing residual carbon [30] and in C–Fe deposits [5] at very high currents. This deposit is in the *diffusion enhanced regime* (DER) (see Sect. 1.5.3). If the supply of gas molecules for deposition is exclusively by direct adsorption from the gas phase, the whole area under the beam spot has the same probability to be deposited upon, resulting in a constant-in-height structure. However, in the DER, the area under the beam is quickly emptied of mobile precursor molecules, due to the large current used and the low gas injection rate. Thus, the supply of new molecules mainly comes from outside the irradiated area, by surface diffusion, with most of the molecules being deposited before reaching the central part of the beam spot. As a consequence, the deposition rate is higher at the edges of the deposit than at the centre. For the lower beam currents used (6.3, 1.6 and 0.4 nA), this effect is not important, resulting in an approximately constant height across the structure. In this case, the regime for deposition is called *precursor-limited regime* (PLR), since the use of relatively high currents (lower than in the DER) with a moderate gas pressure results in a deposition induced by a direct supply from the gas phase. Structures deposited with $I_{BEAM} = 6.3$ nA and $I_{BEAM} = 1.6$ nA are shown in Fig. 7.10b, d. The wire shown in Fig. 7.10c was grown with $I_{BEAM} = 0.4$ nA by a single-pixel-line process (for the other structures shown in Fig. 7.10, the beam was rastered over a rectangular area).

The structural profiles along the y-direction, were fitted to:

$$z_1(y) = z_0 \quad \text{for } |y| < y_1 \tag{7.3.a}$$

$$z_2(y) = A\exp\left[\frac{-(y - y_1)^2}{2\sigma^2}\right] \quad \text{for } y_1 < |y| < y_1 + \Delta \tag{7.3.b}$$

$$z_3(y) = B\exp\left(\frac{-|y|}{y^*}\right) \quad \text{for } |y| > y_1 + \Delta \tag{7.3.c}$$

Equation 7.3.a gives account for the constant height in the centre of the structure when fabricating wide structures, as a consequence of the beam overlap (see for instance Fig. 7.10b, d). The higher thickness at the edges of the structure fabricated in the DER is therefore not considered. The mean roughness for the central part was found to be typically around 10 nm. Gaussian fits of the sides of the structures using Eq. 7.3.b are shown in Fig. 7.10 (red dashed lines), where σ is the standard deviation. Their FWHM ($\approx 2.35\ \sigma$) depends on the beam current, with a value of ~ 500 nm for 24, 6.3 and 1.6 nA, and ~ 150 nm for 0.4 nA. This is several times higher than the FWHM of the Gaussian primary beams used, and is caused by the generation of SE in the substrate in a larger volume than the one targeted by the primary beam [13, 29]. Monte-Carlo simulations evidence this effect [31–33]. The exponential part of the profile (Eq. 7.3c) and the solid blue lines in Fig. 7.10) correspond to the halo deposit mentioned in the previous section, consequence of the generation of SE by BSE (usually called SE^{II}), when irradiating the substrate [7]. The attenuation length y* was found to be of the order of 100–200 nm. In a first approximation, this parameter describes how the probability of SE to escape from the substrate decays exponentially with the distance, from the point where the SE is generated [31, 33]. In the case of the single-pixel-line wires (Fig. 7.10c), the total extension of the deposit should be of the order of the Bethe stop range, i.e. the maximum length an electron travels in the substrate when it does not escape from the sample. We find a value around 1 μm.

Lower currents than those used here would have resulted in a crossover to the *electron-limited regime* (ELR), where the replenishment rate of gas molecules is higher than their dissociation rate. The profiles in this regime are expected to be sharper [13, 29, 31], but we intentionally did not use small currents since the purity of the FEBID-Co is substantially lower for low beam currents (Sect. 7.3).

7.5.3 Magnetization Hysteresis Loops

The room-temperature magnetic properties of each individual structure were analyzed using the high sensitivity spatially-resolved MOKE magnetometer detailed in Sect. 2.4 in the longitudinal configuration. The laser spot (diameter 5 μm) was positioned on top of each single wire. The structures were fabricated far apart from each other in order to measure individual magnetic objects. We measured the magnetic switching along the principal axis of the lines (x-axis), and by a 90° rotation of the substrate, along the axis perpendicular to it (y-axis), while magnetic field sweeps of up to 400 Oe amplitude were applied, at a frequency of 11 Hz. Magnetic saturation was only achieved along the y-direction for AR < 4. This indicates that, at remanence, the magnetostatic energy is the main source of anisotropy for high AR, aligning most of the magnetization along the x-direction. The polycrystalline structure of the FEBID-Co favors this effect, in sharp contrast to single-crystal Co NWs, where the magnetocrystalline anisotropy has a major influence [34]. This result was previously evidenced by magnetotransport measurements on NWs of similar dimensions (Sect. 7.4.1). We obtained a high MOKE signal for each individual structure, including the smallest ones. Only 5% of the samples did not give appreciable Kerr signal. This could be simply caused by a technical deficiency during the growth process, such as the presence of some inhomogeneity in the gas precursor. The measurements shown in this section were obtained after averaging a few loops (10–30 cycles) for a good signal-to-noise ratio.

Some examples of room temperature hysteresis loops measured on different deposits are shown in Fig. 7.11. The blue solid line shows the loop along the principal axis of the structure (x); the red dashed line in Fig. 7.11a shows the loop

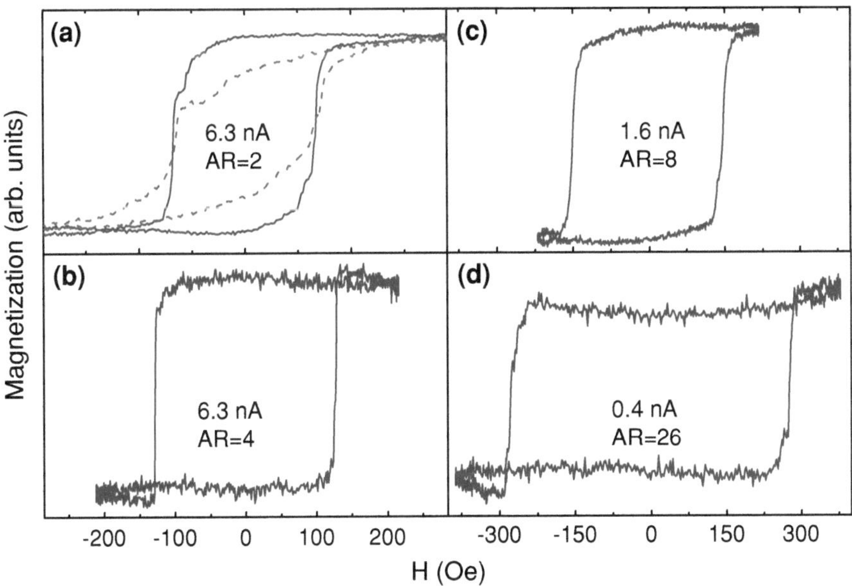

Fig. 7.11 Examples of MOKE loops for different wires (*solid lines*: along the easy x magnetic axis); the *dashed line* in (a) shows the hard y axis loop. The beam current used and the AR of the wires are indicated

measured in the perpendicular direction (y) for a structure with an AR of 2. The order, (a) to (d), corresponds to a progressive increase in the aspect ratio of the wires (from 2 up to 26). For these structures, where the AR > 1, the loops measured along the principal axis of the structure are approximately square-shaped, indicating that shape anisotropy causes the x-direction to be the easy axis of magnetization. Besides, the switching of the magnetization occurs in a field interval of a few tens of Oersted. This has been reported to be a typical feature for the switching in cobalt wires mediated by nucleation of domains followed by DW propagation [19, 20]. Multidomain switching would result in a more progressive reversal of the magnetization [19, 20]. The thickness of our wires is above the 15 nm limit found in reference 288 for single-domain behavior of cobalt wires, so the high AR wires are expected to be monodomain. The reversal of the magnetization along the short axis of the structure, shown in Fig. 7.11a, is a typical hard magnetic axis loop, showing a progressive rotation of M, followed by an abrupt switch.

In order to understand the importance of the halo in the magnetism of the wires, we measured hysteresis loops by focusing the laser spot on that zone. We did not measure a relevant magnetic signal, indicating the low magnetic character of this parasitic deposit. This is likely due to its reduced thickness (AFM images, Fig. 7.10, showing the halo deposit is less than 5 nm at its thickest), making it more susceptible to complete oxidation, together with its significant higher carbon + oxygen/cobalt ratio than in the main deposit, as preliminary Energy Loss Spectroscopy measurements indicate (not shown here). Thus, we conclude that although SEM images show a total structure extending for a few microns in the x–y plane, the structure is FM only in the main wire. This result cannot be adopted as general for FEBID processes, since for instance, a higher iron concentration in the halo with respect to the main deposit has been reported using $Fe(CO)_5$ as gas precursor [11].

The value of the coercive field H_c was determined as the position of the maximum of a Gaussian fit of the derivative of the hysteresis loop. H_c is plotted as a function of the wire width in Fig. 7.12, showing the dramatic effect of the sample geometry on the reversal of the magnetization. The coercivity decreases with the width and increases with the thickness. This trend has been previously observed in wires of different materials [19, 20, 24, 25], and can be understood in magnetostatic terms, since the demagnetizing field depends in a first order approximation on the t/w ratio. By fitting H_c to a function of the form

$$H_c(w) = H_\infty + \frac{\beta}{w} \tag{7.4}$$

where H_∞ and β are free fitting parameters (lines in Fig. 7.12), we obtain a coercive field for an infinitely-wide element $H_\infty = 89 \pm 5$ (70 ± 7) Oe for the 200 (50) nm thick wires, which is comparable to values found in the literature for cobalt polycrystalline thin films [20, 25]. The value w_c for which the above expression of $H_c(w_c)$ equals the coherent rotation switching field value, $H_c(w = w_c) = 2\pi Ms \sim 8350\,Oe$

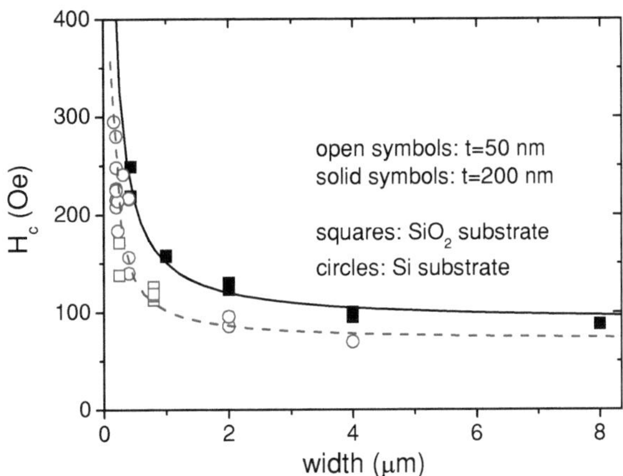

Fig. 7.12 Coercive field as a function of wire width, for the two wire thickness studied. t/w fits to the data are shown (*dashed lines*: t = 50 nm, *solid lines* t = 200 nm)

(M_s value determined by magnetotransport measurements, see Sect. 7.4.1) is 3.5 and 7 nm, for thicknesses of 50 and 200 nm. The values obtained for w_c are an estimation of the wire width below which magnetization reversal would take place by uniform rotation [35], and are of the order of the exchange length (~ 5 nm for Co [25]). These results are in perfect agreement with Refs. [19, 25], where Co wires fabricated by EBL were analyzed in a similar manner.

The critical nucleus responsible for reversal can be approximated to a flat ellipsoid (assuming constant magnetization through the thickness) with a width equal to the width of the wire (w), and a finite length a (a $\leq$ L). Assuming coherent rotation in this nucleus, and applying a Stoner-Wolfharth analysis, Eq. 7.4 can be written as [36]

$$H_c(w, t) = H_\infty + 8M_s \frac{\left(\eta - \frac{1}{\eta}\right)}{\sqrt{1 + \eta^2}} \frac{t}{w} \tag{7.5}$$

where $\eta = a/w$ is the in-plane aspect ratio of the ellipsoid. From our results, we find $\eta = 1 \pm 0.25$ for both the 50 nm and the 200 thick structures. Thus, the volume involved in the initial reversal of M is different from the whole structure, and of the order of a w-long section of the w-wide NW. The reversal process is dominated by the formation of magnetic domains at the edges of the wire (as will be shown in Sect. 7.5.4) or around defects, and subsequent propagation of DWs through the whole wire. The minimization of the magnetostatic energy, by avoiding surface magnetic charges, causes this behavior. The 'effective' nucleation volume appears to be a localized surface-dominated mode, roughly independent of the element size [37]. Our estimation of the nucleation area agrees with values reported for Co NWs fabricated by EBL [25], showing that the magnetization reversal mechanism is similar in FEBID-Co wires, and in Co wires fabricated with more conventional nanolithography techniques.

7.5.4 Micromagnetic Study in FEBID-Co NWs

7.5.4.1 Introduction to Micromagnetic Simulations

Micromagnetic simulations are a theoretical approach which permits the understanding of the magnetic behavior of magnetic nanostructures. The magnetic element simulated is divided into a Cartesian array of parallelepiped cells, and a magnetization vector at the center of each cell (M_i) is defined (with $|M_i| = M_s$). Thus, a "macro-spin" is defined at a region of size of the order of the exchange length. The equilibrium distribution of magnetization for a given value of the applied magnetic field is found by numerically integrating the coupled Landau–Lifschitz–Gilbert equations [38] of each discrete cell (denoted by i), that describes semiclassically the magnetization dynamics:

$$\frac{dM_i}{dt} = -\gamma \, M_i \times \left[H_{\mathit{eff}}\right]^i + \frac{\alpha}{M_s} M_i \times \frac{dM_i}{dt} \qquad (7.6)$$

where γ is the electron gyromagnetic ratio, α is a material-dependent damping coefficient, and H_{eff} is the effective magnetic field, defined as $H_{\mathrm{eff}} = -\mu_0^{-1}\, \partial E/\partial M$, with E the average energy density, which includes all relevant sources of magnetic field, such as exchange, crystalline anisotropy, demagnetization, and Zeeman energy terms, which are evaluated within each particular case.

We have performed micromagnetic calculations using the open access 3-dimensional OOMMF software (the OOMMF code is available at http://math.nist.gov/oommf/). The simulations were done quasistatically, with a damping coefficient $\alpha = 0.5$ to speed up the calculations. The equilibrium points of the system were considered to be reached when dm/dt < 0.1 deg/ns.

7.5.4.2 Micromagnetic Study

Simulations of the switching fields in cobalt polycrystalline wires created by different lithography techniques have already been reported, with a good agreement with experiments [20, 39]. In this work, we have explored the influence of an important aspect related to the FEBID technique on the reversal of the magnetization: a possible gradient in composition through the wire profile when depositing at low beam currents. Because of the complexity of the deposition process, the composition and microstructure can change along the profile, with zones richer in the metallic element and different grain sizes. In Ref. [8], Utke et al. performed a detailed study of Co tips grown with the same gas precursor. They found a complex microstructure and composition, both highly dependent on the beam current. As the precursor decomposition is favored by temperature, the metallic content was higher where the heating was expected to be more important, mainly because of the electron beam irradiation of the substrate. When using low beam currents, the (initial) bottom part of the deposit was found to be lower in metal content than the upper parts, since the heat flow dissipation would change during growth from a 3-dimensional regime to a pseudo-2-dimensional regime. Such a reduced metallic content at the interface between the substrate and the deposit has also been observed in Pt–C deposits grown by FEBID and FIBID (see Sect. 5.1.3.1). Besides, the external parts of the tip had a higher concentration of carbon than the central one.

We have performed micromagnetic calculations using the geometry of the NW shown in Fig. 7.10c (NW grown with $I_{BEAM} = 0.4$ nA, with an expected atomic Co purity of around 85%, see Table 7.1). The structural profile of this NW, measured by AFM, can be described by $z_2(y)$ and $z_3(y)$ (Eqs. 7.3.b and 7.3c), with $y_1 = 0$, $A \approx 40$ nm, $\sigma \approx 64$ nm, $B \approx 18$ nm, $y^* \approx 190$ nm and $\Delta = 112$ nm. The NW was grown with the lowest current used, where a gradient in composition is more likely to occur than at high beam currents, where a purity of almost 100% is expected. Thus, this NW will be mainly composed of pure polycrystalline Co, with some zones composed of a Co_xC_{1-x} mixture, the halo being the zone where x is expected to be minimum (below 20% [11]). Intermediate regions between both zones will have 20% < x < 100%. For simplicity, we have simulated an abrupt gradient in composition in the profile, considering either pure Co zones or non-magnetic zones. This approximation assumes that the magnetic properties of the structures with a considerable C concentration are significantly degraded. The magnetic properties of Co_xC_{1-x} composites deposited with different methods have already been studied [40–42]; for $x \approx 50\%$ the saturation magnetization has been found up to a 5 factor smaller than in bulk Co [41, 42]. Our approximation is also justified by the fact that a mixture of C and metal is substantially more prone to oxidation than a pure metallic layer (Sect. 5.1.3.2). In the magnetic zones, we used the parameters for polycrystalline cobalt: $M_s = 1330$ emu/cm^3, $A = 10 \times 10^{-12}$ J/m, cell size $= 4 \times 4 \times 4$ nm^3. The magnetocrystalline anisotropy K was set to zero, considering that the random orientation of the grains (2–5 nm evidenced by transmission electron microscopy measurements, Sect. 7.3) averages out the magnetocrystalline anisotropy [39, 43]. The importance of a local K in cobalt for small dimensions [44] was not considered for simplicity. The length of the simulated wires was fixed to 2 μm, half of the real size, to speed up the simulations, but large enough to ensure that the shape anisotropy was dominant. The parameters of the profiles studied are detailed in Table 7.2.

Table 7.2 Parameters of the cobalt profiles simulated

	Flat central part		Gaussian edge			Exponential tails		$z_{(y=0)}$ (nm)	$y_{(z=0)}$ (nm)	Switching field (Oe) (± 10 Oe)
	z_0 (nm)	y_1 (nm)	A (nm)	σ (nm)	Δ (nm)	y^* (nm)	B (nm)			
Prof_1	–	0	40	64	112	190	18	40	400	190
Prof_2	–	0	40	64	112	–	–	40	224	210
Prof_3	–	0	40	64	64	–	–	40	128	450
Prof_4	10	75	–	–	–	–	–	40	150	330
Prof_5	–	0	40	64	112	–	–	32	224	310

The length of the wires was fixed to 2 μm in all cases

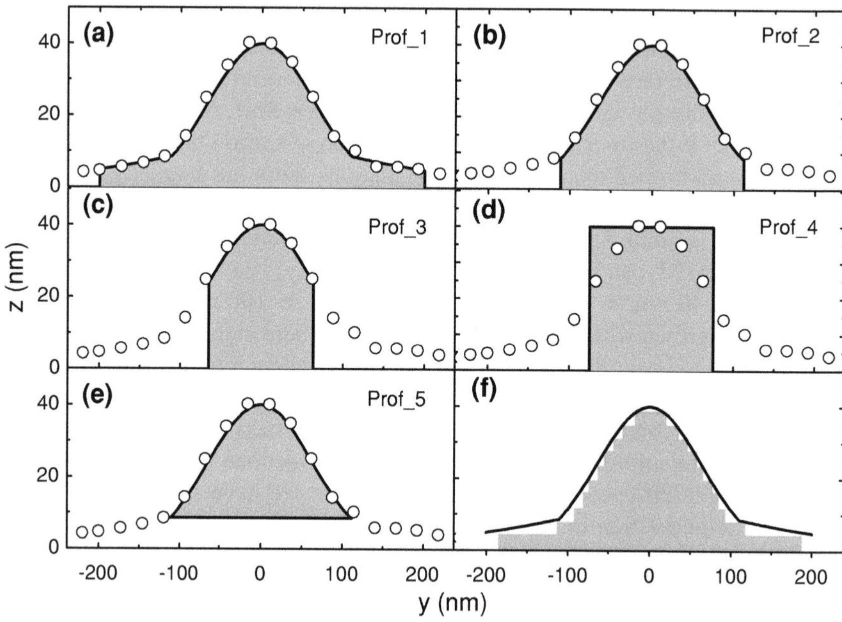

Fig. 7.13 **a–e** Cross-sectional profile of the simulated NWs (*grey area under the curve*) together with the real structural profile (*open circles*). The parameters of the five profiles studied are detailed in Table 7.2. **f** Prof_1 together with its associated discretized version used in the simulations

Figure 7.13 shows the real profile (circles) together with the different profiles simulated (grey area). Prof_1 (Fig. 7.13a) is the real structural profile, considered to be 100% pure polycrystalline Co. Prof_2 (Fig. 7.13b) and Prof_3 (Fig. 7.13c) simulate a lower Co content in the external thinner zones of the NW. In Prof_4 (Fig. 7.13d) we study a perfectly rectangular NW, with width equal to the FWHM of the real structural profile. Finally, Prof _5 (Fig. 7.13e) simulates a NW with low metallic purity for the bottom 8 nm of deposit, at the interface with the substrate.

Figure 7.13f shows Prof_1, together with the corresponding discretized structure used for simulations. The structures are initially forced in a fully saturated state along the long axis, and left to relax under zero field in order to find the remanent state. For all the profiles simulated, the remanent configuration is a monodomain pointing along the main axis of the wire, except near the vertical edges, where the magnetization bends in order to minimize the energy of surface charges. A similar configuration has been observed by magnetic force microscopy measurements in Co polycrystalline NWs of similar dimensions fabricated by EBL [20].

A typical example of the micromagnetic configuration at remanence is shown in Fig. 7.14, in the case of Prof_2, where the micromagnetic configuration in the bottom (centered around $z = 2$ nm), middle ($z = 18$ nm) and top ($z = 38$ nm) planes are shown for one end of the NW. The reversal in all the NWs takes place by DW propagation, in a field interval smaller than the 20 Oe field steps used in

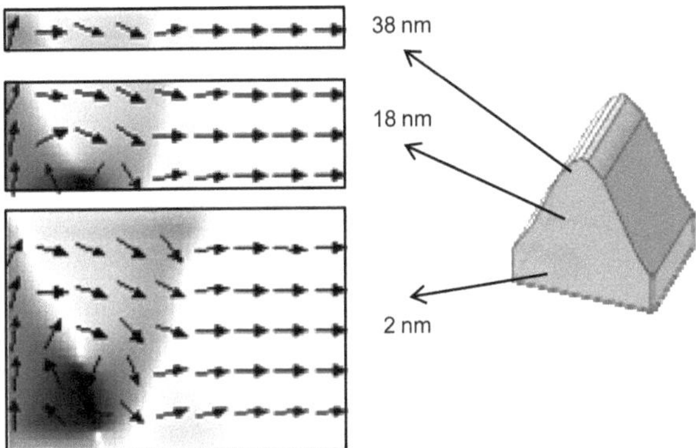

Fig. 7.14 Micromagnetic configuration of one end of Prof_2 wire at remanence. The *top*, *medium* and *bottom pictures* correspond to the micromagnetic configuration at planes centered at $z = 38$, 18 and 2 nm, respectively. The contrast shows the x component of the magnetization, going from *black* $(-M_s)$ to *white* (M_s). The NW has a monodomain state, except for the edges

the simulations. This is in agreement with the relative sharpness of the experimental magnetization loops.

Table 7.2 shows the switching fields obtained for the different profiles. As micromagnetic simulations are done at 0 K and experiments at room temperature, and taking into account the fact that DW motion is a thermally activated process, an increase in the switching fields of around 40–60% from room temperature to 0 K can be expected [22]. This is in good agreement with magnetotransport measurements shown in Sect. 7.4.1. The $T = 0$ extrapolation of the experimentally measured switching value for this NW ($H_{real}^{RT} = 280$ Oe (see Fig. 7.11d) is likely to lie between 400 and 450 Oe. The switching fields obtained from simulations depend on the profile selected, ranging between 190 ± 10 Oe for Prof_1 and 450 ± 10 Oe for Prof_3. Prof_3 gives the switching field value closest to the extrapolated experimental one. This profile has a significantly smaller magnetic volume than the structural one, with a central Gaussian part composed of pure cobalt (bottom width equal to 2σ), and the rest of the real profile being non-magnetic. This result qualitatively agrees with the compositional analysis of reference [8], and could be understood by a low magnetism in the C-rich, highly oxidized thin parts of the profile NW.

7.6 Domain Wall Conduit Behavior in FEBID-Co

7.6.1 Domain Wall Conduit: The Concept

Domain wall conduit behavior is the possibility to displace DWs at much lower fields than those needed to nucleate new domains [45]. If this occurs, DWs can be

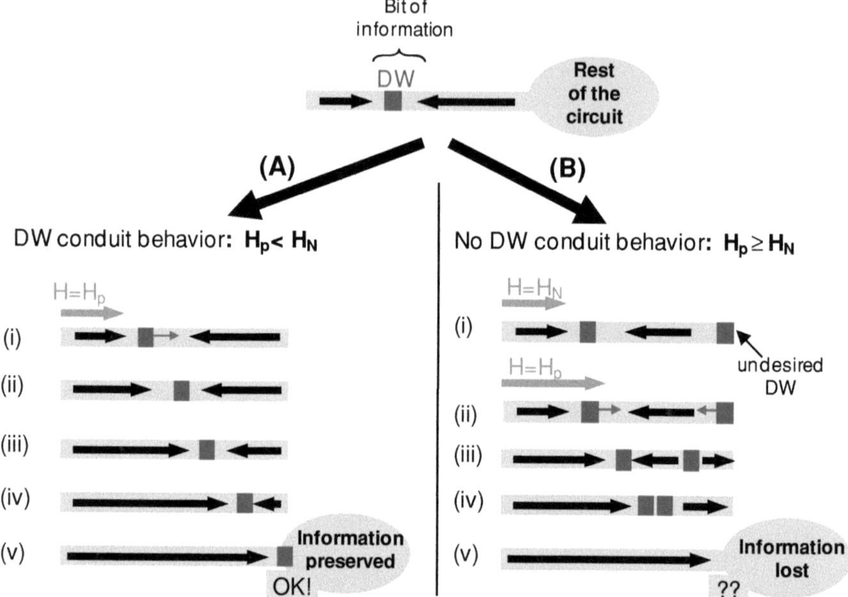

Fig. 7.15 Sketch of the concept of domain wall conduit behavior. The information stored, in the form of a DW, is preserved (*A*) or lost (*B*), depending if there exists good conduit behavior (*A*), or not (*B*). In (*A*) the DW is transmitted to the part of the circuit where it is needed when $H = H_p$. On the contrary, in (*B*) new DWs are created at $H = H_N$ (i-right), and therefore annihilate the pre-existing DW (iv-right) when propagated, at $H = H_P$

treated as particles moved in a controlled way along magnetic networks [46], by the application of an external magnetic field (or by a current, by means of the spin torque effect [47]). In Fig. 7.15 this concept is schematically explained.

This behavior can only occur in the nanoscale, where single-domain structures are feasible, since the magnetostatic energy is the dominating term controlling the magnetic configuration of the nanostructure. DW conduit has been exploited in the last years in planar permalloy NWs, where interesting applications have been proposed (see Sect. 1.7.3)

7.6.2 Creation of L-Shaped Nanowires

For this study, L-shaped NWs were deposited with an electron beam energy of 10 kV, and a beam current of 2.1 nA. For the creation of these structures, we used bitmaps which determine where the beam is scanned. The thickness of the wires was fixed to be around 50 nm, with varying widths (w), from 600 to 150 nm. In Fig. 7.16a, a typical bitmap used is shown. The curvature radio (R) of the L was chosen to be R = 5w, to facilitate the good movement of the walls. Both ends of

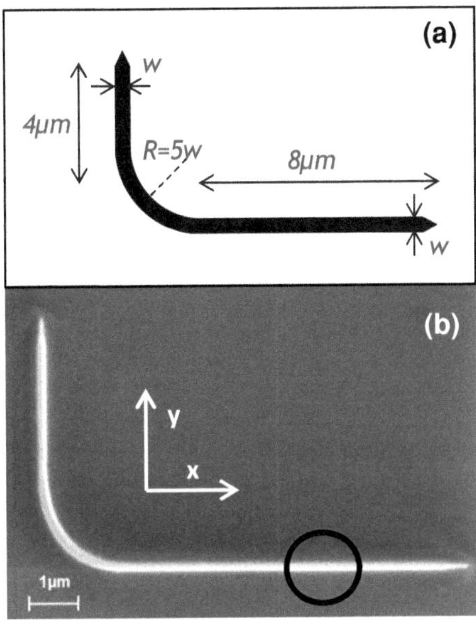

Fig. 7.16 L-shaped nanowires. **a** Bitmap used for the creation of the structures. The *black color* is the zone inside the field of view where the beam is scanned, getting blanked for the white part. **b** SEM image of a 200 nm wide L-shaped cobalt nanowire. The axis nomenclature is maintained in all the section. The *circle* represents the MOKE laser spot (not to scale)

the NW are pointed to avoid domain nucleation [48, 49]. A 200 nm wide structure is shown in Fig. 7.16b. The time taken to fabricate this structure under the chosen conditions is approximately 2 min.

AFM measurements were performed to study the uniformity of the L-shaped wires grown. Scans along different parts of the L-NWs (see dashed lines in Fig. 7.17a) permit to see that a good uniformity in the shape of the structures is attained, with deviations in the section of the wires smaller than 10%. As an example, different scans along a 590 nm wide wire are shown in Fig. 7.17b.

7.6.3 MOKE Measurements and Field Routines

Each structure was analyzed using the high sensitivity MOKE magnetometer. The $\sim$5 µm diameter focused laser spot was placed on the horizontal arm of the wire (see circle in Fig. 7.16b). The component of the magnetization along the x axis was determined by measuring the longitudinal Kerr effect. A typical hysteresis loop is shown in Fig. 7.18, corresponding to a single-sweep experiment. The high signal obtained in only one loop indicates two things: first, this MOKE setup had demonstrated a high sensitivity for permalloy, around 10^{-12} emu (Ref. [12]). The signal to noise ratio is also very good in these experiments on Co NWs, allowing the clear observation of magnetic switching even in single-shot measurements. A rough estimation of the magnetic moment probed in this experiment can be done, taking into account the value for M_s determined from magnetotransport measurements (see Sect. 7.4.1) corresponding to around 10^{-10} emu. Second, it is

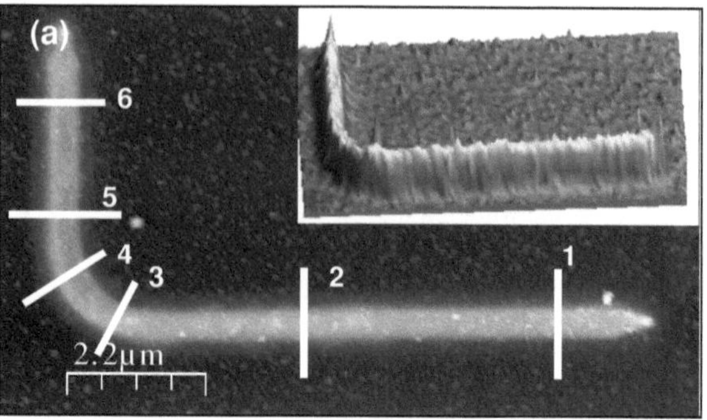

Fig. 7.17 AFM measurements of a 590 nm wide NW. **a** AFM image of the wire, where the *white lines* represent the lines scans performed to determine the profile of the structures in different zones. The *inset* shows the 3D-image. **b** Profiles of the wire, showing good uniformity along the structure. *Lines* are guides to the eye

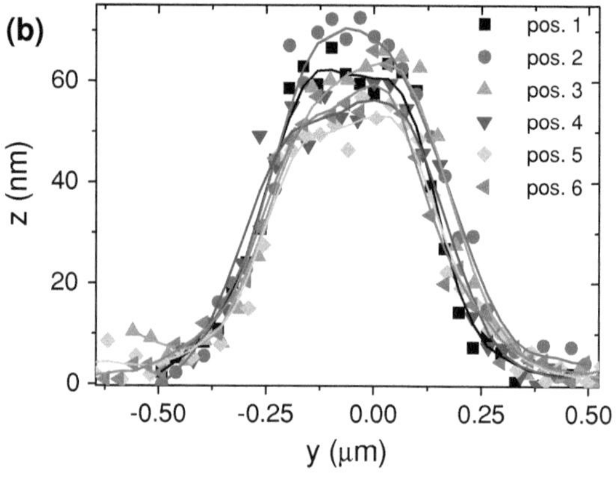

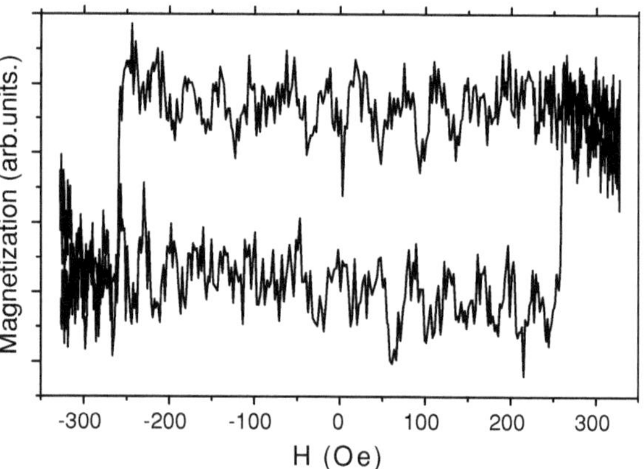

Fig. 7.18 Hysteresis loop of a 200 nm wide nanowire obtained from a single shot measurement. The relevant signal-to-noise ratio provides evidence the high purity of FEBID-Co

another evident proof of the high Co purity of the wires grown under these conditions.

To study the possible good control of DWs of the FEBID-Co, a quadrupole electromagnet was used to apply two types of (H_x, H_y) magnetic field sequences at a frequency of 1 Hz [50–52]. In the first procedure, a DW is initially formed at the

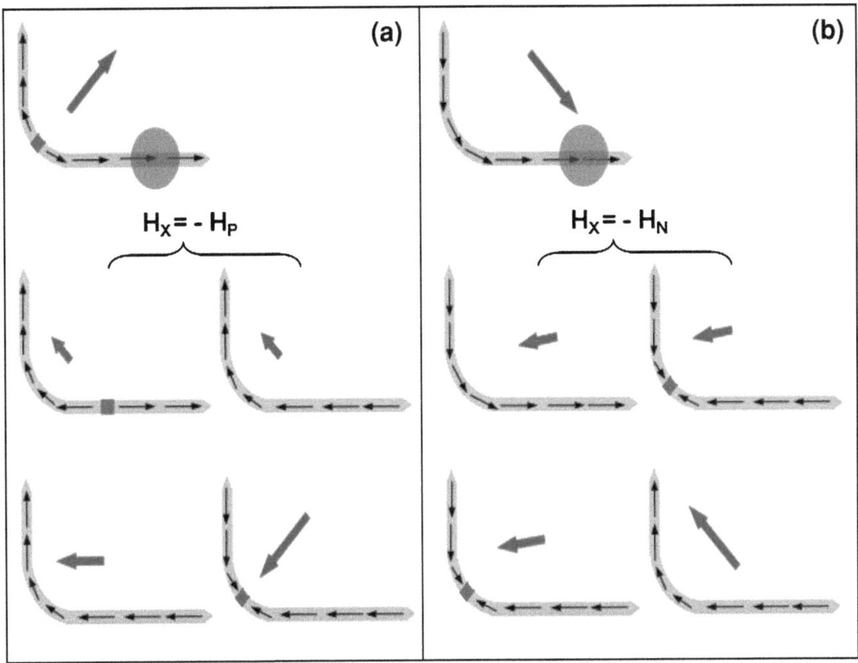

Fig. 7.19 Illustration of the magnetization configurations obtained during half a field cycle (the other half cycle is identical, but with all fields reversed). *External arrows* show the applied magnetic field; *internal arrows* show the direction of magnetization in the nanostructure. The *circle* (not to scale) represents the MOKE laser spot. In (**a**) (propagation) the magnetic field generates a DW at the corner of the L-shaped NW. As H_x changes to negative values, this DW is propagated through the horizontal arm, switching its magnetization at $-H_P$. The subsequent change of H_y from positive to negative values again induces the generation of a DW in the corner. In (**b**) (nucleation), the external field saturates the magnetization along the NW. As H_x is reversed to negative values, a reversed domain is nucleated when $H_x = -H_N$; the magnetization in the horizontal arm switches and a DW is now present at the corner of the NW. The change of H_y to positive values again saturates the magnetization along the NW, in the opposite direction to the initial configuration

corner of the NW. In this case, magnetization reversal occurs by the propagation of this DW in the NW. In the second field sequence, no DW is formed at the corner, so the nucleation of a DW is required to reverse the magnetization. A schematic of the processes occurring in the L-shaped structure in half a cycle (from positive to negative magnetic fields) is shown in Fig. 7.19a, b for the two types of measurements. The other half of the cycle is symmetric.

Hysteresis loops for two particular cases: $w_1 = 500$ nm, and $w_2 = 200$ nm, are shown in Fig. 7.20. (we will cite them as 1 and 2, respectively). To reduce the noise, 30 loops were averaged. In Fig. 7.20a$_1$, a$_2$, the DW initially formed at the corner is propagated through the horizontal NW, switching its magnetization, when the so-called propagation field value ($H_{P1} = 30$ Oe, $H_{P2} = 80$ Oe), is reached. On the contrary, in Fig. 7.20b$_1$, b$_2$ no DW is previously created at the

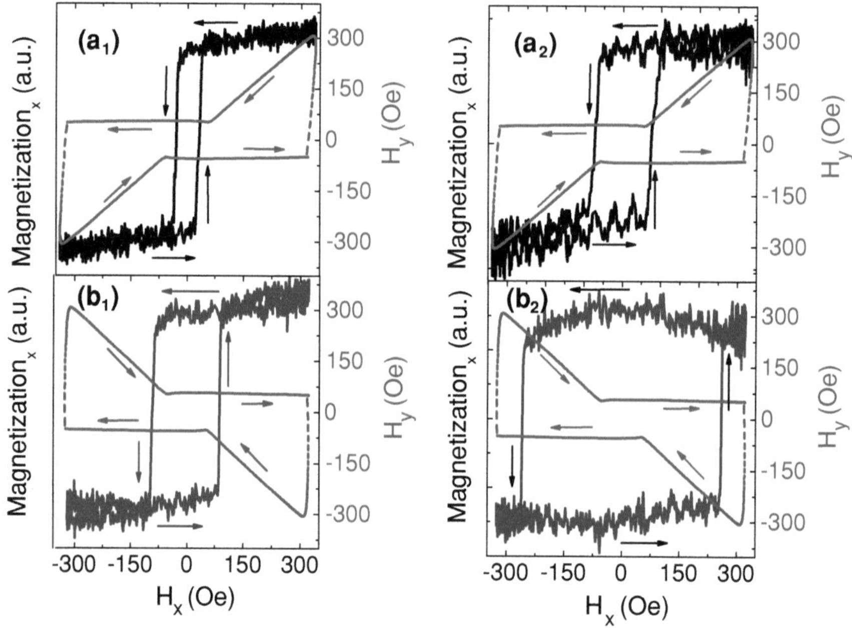

Fig. 7.20 MOKE determined magnetization hysteresis loops (*solid lines*) and corresponding field sequences (*dashed lines*) used to measure (**a**) the propagation and (**b**) the nucleation field for two NWs, with widths of (**1**) 500 nm and (**2**) 200 nm. The *arrows* on the graphs indicate the sense of variation of the magnetization and the field as the routine is applied

corner. The reversal in this case happens at significant higher magnetic fields (at the nucleation field $H_{N1} = 91$ Oe, $H_{N2} = 259$ Oe). In both NWs, H_P and H_N are very different, with $H_P < H_N$. This behavior is fundamental for possible applications of this material to devices which rely on the formation and control of DWs. It must be noticed that a small field offset in the y-direction was maintained in both sequences after the initialization of the magnetization, ensuring that the DW remains in the bottom of the structure and does not propagate in the y direction [50, 51]. This offset (H_y^{off}) was chosen so that $H_y^{off} > H_p$ for the wider NWs, and was maintained when measuring all the NWs for consistency. Systematic measurements of H_p and H_N as a function of H_y show that this transverse offset has very little effect on the switching fields (see Fig. 7.21): a maximum variation of 15% was observed in H_p for a 150 nm wire when H_y varied between 20 and 80 Oe. The error bars on Fig. 7.22 take into account this small effect.

In Fig. 7.22 the propagation and nucleation fields are shown as a function of the NW width for all the wires studied. The dependence of H_N, inverse with the width, is the same as discussed in Sect. 7.5.3, understood in magnetostatic terms. These nucleation fields are slightly higher than those measured for rectangular wires, due to the pointing ends. We can see that H_P has a similar dependence on width. A good fit is also obtained with the inverse of w.

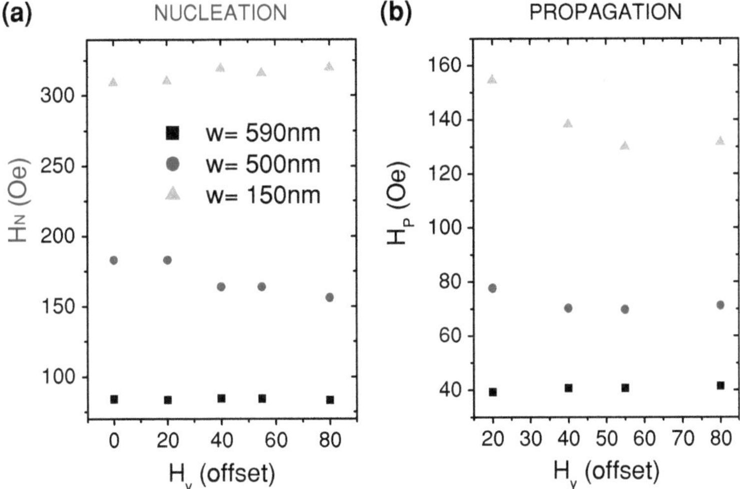

Fig. 7.21 Effect of a H_y-offset in **a** the nucleation and **b** propagation fields

Fig. 7.22 Propagation (H_P) and nucleation fields (H_N) as a function of the nanowire width. The error bars for the fields take into account the effect of the small H_y-offset used in the field sequences (see text for details)

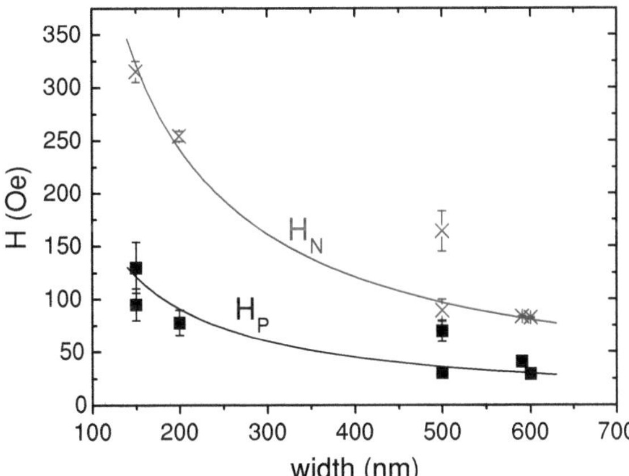

In the case of the nucleation loops, abrupt transitions are always observed, suggesting that the reversal mechanism is by the nucleation of domains plus propagation, starting from a monodomain state (Sect. 7.5.3). In the case of DW propagation, the transition is not as abrupt as in nucleation field measurements, especially for the narrow NWs. This widening could be thought to be caused by a series of pinning sites under the MOKE spot. However, single-shot loops (not shown here) are all sharp, indicating this is not the case. Therefore, the relatively wider transitions for H_p are caused by the statistical averaging of abrupt switches occurring at close but slightly different fields. These results show that, for the dimensions chosen, the conduction of domain walls in FEBID-Co NWs is excellent. The propagation fields are of a few tens of oersted for the wider NWs. These values are comparable with those ones obtained for thin and narrow permalloy NWs used nowadays by the scientific community in the area [48, 50–52]. This is not the case for these narrower NWs, were H_p are too high for most practical magnetoelectronic applications. Thinner NWs should be studied to determine if good DW control is achieved for narrow FEBID-Co NWs in a practical range of fields.

7.7 Conclusions

We have done a complete characterization of the compositional, microstructural and magnetotransport properties of FEBID-Co nanowires. EDX measurements show Co percentages higher than 90% in deposits grown under high beam currents, with percentages around 80% for low currents. We explain the evolution of composition, as well as the high purity attained, in terms of the high local heating induced by the Field-Emission Gun Beam. HRTEM images show that FEBID-Co is polycrystalline, with grain sizes ~ 2–5 nm. For high currents, Hall, planar Hall effect and MR are the typical ones for polycrystalline cobalt, with metallic conduction. From these measurements we infer a saturation magnetization very close to the bulk value. NWs grown under low currents show degraded magnetotransport properties.

We report direct hysteresis loop measurements performed by MOKE on single Co wires fabricated by FEBID. We have done a systematic study of the reversal of the magnetization in cobalt wires grown by FEBID in a wide range of dimensions. The magneto-optical signal of the grown structures is high for all beam currents, giving account of the magnetic character of the FEBID-Co under the chosen conditions. The coercive field is proportional to the thickness/width ratio, indicating that the anisotropy is mainly due to the shape of the structure. For high aspect ratios, the shape anisotropy forces the magnetization to lie in the longitudinal direction of the NW in remanence.

We showed that the effective nucleation area is not the full wire, but rather a small section of length of the order of the width of the wire. Similar results have been published on cobalt and permalloy wires fabricated by electron beam and focused ion beam lithography. This picture agrees with the micromagnetic simulations we have performed, which indicate that high aspect ratio wires are monodomain expect for the ends, where the magnetization follows the edges, in order to minimize the magnetostatic energy.

We studied the influence of a gradient in composition on the switching field, when depositing at the lowest current used. The magnetic profile considered that best reproduces the experimental switching fields is smaller than the structural profile, and made of a central narrow Gaussian structure.

By the creation of L-shaped NWs, and the application of 2-directions in-plane magnetic field routines, we have demonstrated well-controlled formation and motion of DWs in cobalt FEBID NWs. In the range of dimensions studied, the process for reversal of the NW is well controlled by the magnetic fields applied, being possible to select either DW propagation, or nucleation of a new domain. Both kinds of processes occur at well-defined fields. Significant differences are found for the fields necessary to reverse the magnetization, if a DW is initially present, or not, in the nanostructure. This demonstrates the conduit properties of the Co FEBID nanowires.

The excellent results obtained for FEBID-Co, makes this technique very attractive for future applications. Implementation of complex magnetic NW networks, necessary for real applications based on DW control, is a routine task with this technique, with the enormous advantage of the simplicity with respect to others with nanometer resolution, and the possibility to create these devices, in principle, on any surface. Moreover, the good performances of FEBID for creating 3D-structures, makes this technique a unique tool for the fabrication of magnetic nanostructures.

References

1. M. Takeguchi, M. Shimojo, K. Furuya, Fabrication of magnetic nanostructures using electron beam induced chemical vapour deposition. Nanotechnology **16**, 1321 (2005)
2. M. Takeguchi, M. Shimojo, R. Che, K. Furuya, Fabrication of a nano-magnet on a piezo-driven tip in a TEM sample holder. J. Mater. Sci. **41**, 2627 (2006)
3. M. Shimojo, M. Takeguchi, K. Mitsuishi, M. Tanaka, K. Furuya, Mechanisms of crystalline iron oxide formation in electron beam-induced deposition. Jpn. J. Appl. Phys. **9B**, 6247 (2007)
4. W. Zhang, M. Shimojo, M. Takeguchi, R.C. Che, K. Furuya, Generation mechanism and in situ growth behavior of α-iron nanocrystals by electron beam induced deposition. Adv. Eng. Mater. **8**, 711 (2006)
5. M. Beljaars (2009) Electron beam induced deposition of iron. Thesis, Eindhoven University of Technology, The Netherlands
6. I. Utke, P. Hoffmann, R. Berger, L. Scandella, High resolution magnetic force microscopy supertips produced by focused electron beam induced deposition. Appl. Phys. Lett. **80**, 4792 (2002)
7. D. Spoddig, K. Schindler, P. Rodiger, J. Barzola-Quiquia, K. Fritsch, H. Mulders, P. Esquinazi, Transport properties and growth parameters of PdC and WC nanowires prepared in a dual-beam microscope. Nanotechnology **18**, 495202 (2007)
8. I. Utke, J. Michler, P. Gasser, C. Santschi, D. Laub, M. Cantoni, P.A. Buffat, C. Jiao, P. Hoffmann, Cross section investigations of compositions and sub-structures of tips obtained by focused electron beam induced deposition. Adv. Eng. Mater. **7**, 323 (2005)
9. A. Lapicki, E. Ahmad, T. Suzuki, Ion beam induced chemical vapor deposition (IBICVD) of cobalt particles. J. Magn. Magn. Mater. **240**, 1 (2002)
10. I. Utke, T. Bret, D. Laub, Ph. Buffat, L. Scandella, P. Hoffmann, Thermal effects during focused electron beam induced deposition of nanocomposite magnetic-cobalt-containing tips. Microelectron. Eng. **73**, 553 (2004)
11. Y.M. Lau, P.C. Chee, J.T.L. Thong, V. Ng, Properties and applications of cobalt-based material produced by electron-beam-induced deposition. J. Vac. Sci. Technol. A **20**(4), 1295 (2002)
12. D.A. Alwood, G. Xiong, M.D. Cooke, R.P. Cowburn, Magneto-optical Kerr effect analysis of magnetic nanostructures. J. Phys. D Appl. Phys. **36**, 2175 (2003)

13. I. Utke, P. Hoffmann, J. Melngailis, Gas-assisted focused electron beam and ion beam processing and fabrication. J. Vac. Sci. Technol. B **26**, 1197 (2008)
14. G.A. West, K.W. Beeson, Chemical vapor deposition of cobalt silicide. Appl. Phys. Lett. **53**, 740 (1998)
15. J. Kötzler, W. Gil, Anomalous Hall resistivity of cobalt films: evidence for the intrinsic spin-orbit effect. Phys. Rev. B **72**, 060412(R) (2005)
16. S. Onoda, N. Sugimoto, N. Nagaosa, Intrinsic versus extrinsic anomalous Hall effect in ferromagnets. Phys. Rev. Lett. **97**, 126602 (2006)
17. S. Sangiao, L. Morellón, G. Simón, J.M. De Teresa, J.A. Pardo, M.R. Ibarra, Anomalous Hall effect in Fe (001) epitaxial thin films over a wide range in conductivity. Phys. Rev. B **79**, 014431 (2009)
18. M. Rubinstein, F.J. Rachford, W.W. Fuller, G.A. Prinz, Electrical transport properties of thin epitaxially grown iron films. Phys. Rev. B **37**, 8689 (1988)
19. B. Leven, G. Dumpich, Resistance behavior and magnetization reversal analysis of individual Co nanowires. Phys. Rev. B **71**, 064411 (2005)
20. M. Brands, G. Dumpich, Experimental determination of anisotropy and demagnetizing factors of single Co nanowires by magnetoresistance measurements. J. Appl. Phys. **98**, 014309 (2005)
21. W. Gil, D. Görlitz, M. Horisberger, J. Kötzler, Magnetoresistance anisotropy of polycrystalline cobalt films: geometrical-size- and domain-effects. Phys. Rev. B **72**, 134401 (2005)
22. A. Himeno, T. Okuno, K. Mibu, S. Nasu, T. Shinjo, Temperature dependence of depinning fields in submicron magnetic wires with an artificial neck. J. Magn. Magn. Mater. **286**, 167 (2005)
23. W. Wernsdorfer, E. Bonet Orozco, K. Hasselbach, A. Benoit, B. Barbara, N. Demoncy, A. Loiseau, H. Pascard, D. Mailly, Experimental evidence of the Néel-Brown model of magnetization reversal. Phys. Rev. Lett. **78**, 1791 (1997)
24. R.P. Cowburn, Property variation with shape in magnetic nanoelements. J. Phys. D Appl. Phys. **33**, R1 (2000)
25. W. Casey Uhlig, J. Shi, Systematic study of the magnetization reversal in patterned Co and NiFe Nanolines. Appl. Phys. Lett. **84**, 759 (2004)
26. J. Vogel, J. Moritz, O. Fruchart, Nucleation of magnetisation reversal, from nanoparticles to bulk materials. C.R. Physique **7**, 977 (2006)
27. M. Brands, R. Wieser, C. Hassel, D. Hinzke, G. Dumpich, Reversal processes an domain wall pinning in polycrystalline Co-nanowires. Phys. Rev. B. **74**, 174411 (2006)
28. T. Wang, Y. Wang, Y. Fu, T. Hasegawa, F.S. Li, H. Saito, S. Ishio, A magnetic force microscopy study of the magnetic reversal of a single Fe nanowire. Nanotechnology **20**, 105707 (2009)
29. W.F. van Dorp, C.W. Hagen, A critical literature review of focused electron beam induced deposition. J. Appl. Phys. **104**, 081301 (2008)
30. L. Reimer, M. Wächter, Contribution to the contamination problem in transmission electron microscopy. Ultramicroscopy **3**, 169 (1978)
31. D.A. Smith, J.D. Fowlkes, P.D. Rack, A nanoscale three-dimensional Monte Carlo simulation of electron-beam-induced deposition with gas dynamics. Nanotechnology **18**, 265308 (2007)
32. Z.-Q. Liu, K. Mitsuishi, K. Furuya, A dynamic Monte Carlo study of the in situ growth of a substance deposited using electron-beam-induced deposition. Nanotechnology **17**, 3832 (2006)
33. K.T. Kohlmann-von Platen, J. Chlebeck, M. Weiss, K. Reimer, H. Oertel, W.H. Brunger, Resolution limits in electron-beam induced tungsten deposition. J. Vac. Sci. Technol. B **11**, 2219 (1993)
34. Y. Henry, K. Ounadjela, L. Piraux, S. Dubois, J.-M. George, J.-L. Duvail, Magnetic anisotropy and domain patterns in electrodeposited cobalt nanowires. Eur. Phys. J. B **20**, 25 (2001)

35. E.H. Frei, S. Shtrikman, D. Treves, Critical size and nucleation field of ideal ferromagnetic particles. Phys. Rev. **106**, 446 (1957)
36. S.Y. Yuan, H.N. Bertram, J.F. Smyth, S. Schultz, Size effects of switching fields of thin permalloy particles. IEEE Trans. Magn. **28**, 3171 (1992)
37. H. Suhl, H.N. Bertram, Localized surface nucleation of magnetization reversal. J. Appl. Phys. **82**(12), 6128 (1997)
38. L. Landau, E. Lifshitz, On the theory of the dispersion of magnetic permeability in ferromagnetic bodies. Phys. Z. Sowjetunion **8**, 153 (1935)
39. B. Hausmanns, T.P. Krome, G. Dumpich, E.F. Wassermann, D. Hinzke, U. Nowak, K.D. Usadel, Magnetization reversal process in thin Co nanowires. J. Magn. Magn. Mater. **240**, 297 (2002)
40. M. Yu, Y. Liu, D.J. Sellmyer, Structural and magnetic properties of nanocomposite Co:C films. J. Appl. Phys. **85**, 4319 (1999)
41. H. Wang, S.P. Wong, W.Y. Cheung, N. Ke, R.W.M. Kwok, Magnetic properties and structure evolution of amorphous Co-C nanocomposite films prepared by pulsed filtered vacuum arc deposition. J. Appl. Phys. **88**, 4919 (2000)
42. J. Shi, M. Azumi, O. Nittono, Structural and magnetic properties of Co-C composite films and Co/C multilayer films. Appl. Phys. A **73**, 215 (2001)
43. D. Grujicic, B. Pesic, Micromagnetic studies of cobalt microbars fabricated by nanoimprint lithography and electrodeposition. J. Magn. Magn. Mater. **285**, 203 (2005)
44. R.M.H. New, R.F.W. Pease, R.L. White, Switching characteristics of submicron cobalt islands. IEEE Trans. Magn. **6**, 3805 (1995)
45. R.P. Cowburn, D. Petit, Spintronics: turbulence ahead. Nat. Mater. **4**, 721 (2005)
46. T. Ono, H. Miyajima, K. Shigeto, M. Kibu, N. Hosoito, T. Shinjo, Propagation of a magnetic domain wall in a submicrometer magnetic wire. Science **284**, 468 (1999)
47. D.C. Ralph, M.D. Stiles, Spin transfer torques. J. Magn. Magn. Mater **320**, 1190 (2008)
48. D.A. Alwood, G. Xiong, C.C. Faulkner, D. Atkinson, D. Petit, R.P. Cowburn, Magnetic domain-wall logic. Science **309**, 1688 (2005)
49. D. Atkinson, D.S. Eastwood, L.K. Bogart, Controlling domain wall pinning in planar nanowires by selecting domain wall type and its application in a memory concept. Appl. Phys. Lett. **92**, 022510–022511 (2008)
50. D. Petit, A.-V. Jausovec, D.E. Read, R.P. Cowburn, Domain wall pinning and potential landscapes created by constrictions and protusions in ferromagnetic nanowires. J. Appl. Phys. **103**, 11437 (2008)
51. D. Petit, A.-V. Jausovec, H.T. Zeng, E. Lewis, L. O'Brien, D.E. Read, R.P. Cowburn, High efficiency domain wall gate in ferromagnetic nanowires. Appl. Phys. Lett. **93**, 163108 (2008)
52. D. Petit, A.-V. Jausovec, H.T. Zeng, E. Lewis, L. O'Brien, D.E. Read, R.P. Cowburn, Mechanism for domain wall pinning and potential landscape modification by artificially patterned traps in ferromagnetic nanowires. Phys. Rev. B. **79**, 214405 (2009)

Chapter 8
Conclusions and Outlook

The subject of this doctoral thesis is a study of electrical and magnetic properties in several nanometric materials with a different range of scales: thin films, atomic-sized nanoconstrictions, and micro and nanowires. In this chapter the main conclusions of the work are summarized and future perspectives discussed. The general conclusions, especially regarding the use of experimental techniques and methods are described first. Secondly, specific conclusions relating the three general subjects of the thesis are set out: Fe_3O_4 thin films, nanoconstrictions created by FIB and nanowires created by FEBID/FIBID.

8.1 General Conclusions

The doctoral thesis has been completed thanks to the synergy of two research institutes in Zaragoza: the Institute of Materials Science of Aragón (ICMA), an organization working under the aegis of the CSIC and the University of Zaragoza with long and reputed experience in materials science research, and the Institute of Nanoscience of Aragón (INA), pertaining to the University of Zaragoza, an interdisciplinary institute which has state of the art equipment for use in the nanoscience field. This is the first thesis defended in this region where top-down micro and nanolithography techniques employing equipment provided by Zaragoza institutions have been used. Novel methods and procedures have been developed which will serve as useful input for further institutional research into nanoscale materials and devices. I should give special mention to the development of certain photolithography and nanolithography processes which require the learning of a large number of experimental techniques and protocols. Furthermore, the assembly and control of equipment for the study of magnetotransport properties has also been carried out. Finally, some successful procedures for the control

A. Fernandez-Pacheco, *Studies of Nanoconstrictions, Nanowires and Fe3O4 Thin Films*, Springer Theses, DOI: 10.1007/978-3-642-15801-8_8,
© Springer-Verlag Berlin Heidelberg 2011

of the resistance of devices while being used in Dual Beam equipment also represent an innovative feature of this work, even at an international level.

8.2 Fe_3O_4 Epitaxial Thin Films

Due to the high spin polarization of Fe_3O_4, with potential applications in spintronics devices, magnetotransport properties in epitaxial Fe_3O_4 thin films grown on MgO (001) have been thoroughly studied. The photolithography process performed has permitted us to study the longitudinal and transversal components of the resistivity tensor in a wide range of temperatures, above and quite some way below the Verwey transition.

Resistivity and magnetoresistance measurements at moderate magnetic fields reproduce previous results in the literature, supporting the epitaxial character of the films. The transport properties of the ultra-thin films are highly dominated by the scattering produced in the antiphase boundaries. The thinnest 4 nm-thick film, reaches resistivity values of 150 mΩ cm, $\sim$40 times higher than bulk. The Verwey transition gets smoother and shifts to lower temperature with the film thickness, becoming negligible for $t < 10$ nm. The MR dependence on the film thickness is dominated by the density of APBs, evidencing the tendency towards the superparamagnetism of the domains between the boundaries for thicknesses below 40 nm.

The MR has been measured for a 40 nm thick sample up to 30 Tesla as a function of temperature. The MR is not saturated in the high magnetic fields applied, and presents a peak at the Verwey transition of approximately -20% at maximum field. These measurements show that MR in films is caused by the combination of a—temperature and thickness dependent—intrinsic and extrinsic effect. Further work is necessary for the full understanding of this complex scenario.

The high resistivity of the films together with a moderate anisotropic magnetoresistance results in giant values for the planar Hall effect. We have systematically studied this effect, finding a record value at room temperature of -60 $\mu\Omega$ cm for a 5 nm-thick film. Below the Verwey transition, colossal values never attained before in any material, in the mΩ cm range, are measured. The changes in the AMR sign below the transition, as well as the dependence of its magnitude on temperature are not yet well understood.

An important experimental result has been obtained regarding the study of the anomalous Hall effect in four orders of magnitude of the longitudinal conductivity (10^{-2} Ω^{-1} cm$^{-1} < \sigma_{xx} < 10^2$ Ω^{-1} cm^{-1}). The dependence $\sigma_H \propto \sigma_{xx}^{1.6}$ is fulfilled in all cases, with no influence of the mechanism responsible for conduction, different degrees of scattering at the APBs, etc. This is a universal behavior of all magnetite results reported in the literature, as well as of all compounds in the dirty regime of conductivities ($\sigma_{xx} < 10^4$ Ω^{-1} cm^{-1}). In our opinion, deeper theoretical work should be done to clarify how the model predicting this behavior fits with non-metallic conduction.

The ordinary Hall effect has been accurately separated from the anomalous one for 40 nm-thick samples. We derive that ~ 1 electron per f.u. contributes to the conduction at room temperature. Thinner samples have a smaller effective density of conducting electrons, as a consequence of the APBs.

The first magnetic tunnel junction based on Fe$_3$O$_4$ electrodes with an MgO barrier are currently (2009) being grown in our laboratory by Orna. We hope that the experience gained in this thesis on the magnetotransport properties and photolithography process of Fe$_3$O$_4$ films will contribute to the successful development of these challenging devices in Zaragoza.

8.3 Creation of Atomic-sized Constrictions in Metals using a Focused-Ion-Beam

We have developed a novel method to fabricate controlled atomic-sized constrictions in metals, by measuring the conduction of the device while a focused-ion-beam etching process is in progress. We have demonstrated that the technique works when using Cr and Fe, creating stable atomic-sized constrictions under vacuum inside the same chamber. This method has two main advantages with respect to more conventional procedures (STM, MBJ, EBJ, etc.). First, the FIB processes can be introduced into the standard protocols for the integrated circuits industry, so possible applications of these devices would be feasible using this methodology, which is not the case for the others. Second, regarding atomic-sized contacts in magnetic materials, where huge magnetoresistance effects observed in the literature are under debate, we consider that the good adherence of the atoms forming the contact to a substrate can minimize artifacts, making this technique more suitable for this kind of study.

The instability of metallic contacts when exposed to ambient conditions has been evidenced in iron nanoconstrictions. In spite of a nanometric protective covering, the samples get degraded when exposed to the atmosphere, either becoming completely degraded or getting into the tunneling regime of conductivities. We have shown results for one particular sample in this conduction regime. The TMR has a value of 3%, ~ 30 times higher and of different sign than bulk Fe. The TAMR effect is of the order of 2% and with a much more abrupt dependence than the $\cos^2\theta$ diffusive behavior. The effects are of contrary sign that if they were caused by a magnetostriction contribution. These results are not systematic, but evidence the preservation of the magnetic character of the material after ion irradiation. This constitutes a promising field for the study of BMR and BAMR. New strategies for protection of the constrictions, together with the possibility of implementing a device for magnetic measurements attached to the FIB chamber, to avoid the exposure of the sample to air, would be substantial advances for this research. The method developed is now being successfully applied in our laboratory for other types of materials, of interest from fundamental and applied points of view.

8.4 Nanowires Created by Focused Electron/Ion Beam Induced Deposition

An important part of the thesis has been devoted to the study of nanowires created by these promising nanolithography techniques. The work has been done with three types of gas precursors: $(CH_3)_3Pt(CpCH_3)$, $W(CO)_6$ and $Co_2(CO)_8$, and using in the first case both focused electrons and ions to dissociate the precursor molecules, using ions in the second case, and using electrons in the third. In such local deposition, the material is in general composed of a mixture of carbon and metal. This broad-based work provides a general view of the complex physical phenomena under these processes, which produce completely different microstructures in the material. With the $(CH_3)_3Pt(CpCH_3)$ precursor, an amorphous carbon matrix with nanometric platinum clusters embedded inside is produced. With $W(CO)_6$, an amorphous disordered carbon–tungsten microstructure is formed, whereas using $Co_2(CO)_8$, under high currents, almost pure polycrystalline cobalt is grown. Several physical phenomena have been studied in each case, all of them influenced by the particular microstructure of the material.

In the case of Pt–C NWs, we have observed a Mott-Anderson metal–insulator transition as a function of the carbon/metal concentration, with the resistivity changing from 10^8 down to 10^3 $\mu\Omega$ cm. Metallic NWs are obtained for thicknesses above 50 nm. This explains, in principle, contradictory results in the literature up to now. The insulator NWs present an Efros-Shklovskii variable range hopping conduction. The discovery of an exponential decrease in the conductance for the most resistive samples demonstrates the validity of the theories for hopping under strong electric fields in a nanometric structure. This work shows the importance of tuning the parameters in a FEBID/FIBID process, since in this case insulator or metallic NWs can be created. The method for controlling the resistance as the deposit is being performed, analogous to that for constrictions, has been evidenced to be very useful, and is now of general use in our laboratory.

By studying the low temperature electrical properties of FIBID-W micro and nanowires, we have determined that wires with sub-100 nm lateral width retain the superconducting properties. The critical temperatures of the wires are ~ 5 K, with critical fields (Bc$_2$) of ~ 8 Tesla and critical density currents of 0.06–0.09 MA/cm^2. This, together with the outstanding properties of this material as determined by Scanning-Tunneling-Spectroscopy measurements made by Prof. Sebastián Viera's research group, make FIBID-W a promising nanoscale superconductor at Helium liquid temperatures. Nano-SQUIDS and Andreev reflection nanocontacts based on this nanodeposit are promising applications currently being developed by our group.

By the use of moderately high beam currents, FEBID-Co nanowires were obtained, reaching the highest degree metallic purity achieved so far, of $\sim 90\%$. As a result, magnetotransport measurements in Co NWs deposited in these conditions are similar to those of pure Co wires fabricated by conventional nanolithography techniques. The resistivity at room temperature is 40 $\mu\Omega$ cm, only 7 times higher than the bulk value, with fully metallic temperature dependence. The anisotropic magnetoresistance is approximately -0.7%. From Hall effect measurements we derive the saturation magnetization, $M_s = 1330 \pm 20$ emu/cm^3, a 97% of the bulk value. When using low currents, the degradation of these properties was observed ($\rho = 10800$ $\mu\Omega$ cm, semiconducting $\rho(T)$ and AMR approximately -0.1%). By the use of the spatially-resolved longitudinal Magneto-Optical-Kerr-effect, we have studied hysteresis loops in single wires with different aspect ratios (from 1–26, with minimum width of 150 nm for thickness of 40 nm). The coercive field is linear with the thickness/width ratio, indicating that the magnetostatic energy dominates the configuration of the magnetization in the wires. For high aspect ratios, the shape anisotropy forces the remanent magnetization to lie in the longitudinal direction of the NW. The reversal of the magnetization with fields along the wire axis is caused by a nucleation plus propagation process. The effective nucleation area is not the full wire, but rather a small section of length of the order of the width of the wire; micromagnetic simulations confirm this statement. The behavior of these NWs is similar to NWs created by more conventional lithography techniques. By creating L-shaped NWs, and using two-dimensional field routines, we have demonstrated the good domain wall conduit behavior of this material, with propagation and nucleation fields of the order of 150–300 Oe and 50–100 Oe, respectively. The unique advantages of this technique for creating 3-dimensional nanostructures make this result especially interesting. Steps in this direction are in progress. Moreover, further progress should be achieved by performing experiments where the magnetization is switched by a spin-torque effect, and not by an external magnetic field.

Future works should, in general, be devoted to the study of the resolution limits of these techniques, and determine how their functional properties are affected when the sizes are reduced substantially. Another important aspect is to try to obtain purely metallic nanostructures, of interest in nanoelectronics. Further, functionality of these deposits in a wide variety of applications is highly desirable. Finally, the combination of these materials to create hybrids such as magnetic/superconducting devices, together with effective control of the resistance, could allow the study of fascinating phenomena.

Curriculum Vitae

[Status: July 2010]

Name: Dr. Amalio
                   Fernández-Pacheco
                   Chicón

Date and   24/1/1981
Place   in
of Birth: Zaragoza (Spain)

Professional   University of
Address: Cambridge,
                   Department of
                   Physics, JJ Thomson Avenue, Cambridge, CB3 0HE

Email: af457@cam.ac.uk

Personal Webpage: http://cambridge.academia.edu/amalio/

Scientific Background

Current position: Research Associate at the Cavendish laboratory, University of Cambridge (UK)
Jan to May 2010: Research Associate at the Blackett laboratory, Imperial College London (UK)
2009: Ph.D. in Physics, Faculty of Science, University of Zaragoza (Spain)
Title: Electrical conduction and magnetic properties of nanoconstrictions and nanowires created by focused electron/ion beam and of Fe_3O_4 thin films
Supervisors: Prof. M. Ricardo Ibarra and Dr. José María De Teresa

Committee: Nobel Laureate Prof. Albert Fert (CNRS/Thales, Palaiseau, France), Sebastián Vieira (Universidad Autónoma de Madrid, Spain), Jagadeesh Moodera (Bitter Laboratory MIT, USA), Russell P. Cowburn (Imperial College London, United Kingdom) Luis Morellón (University of Zaragoza, Spain)
Mark: Apto *Cum Laude*
2007: Master degree in Physics and Physical Technologies, Faculty of Sciences, University of Zaragoza (Spain)
2000–2005: Graduate degree in Physics with honours, Faculty of Sciences, University of Zaragoza (Spain)

Prizes

- **2010:** Springer Theses award, recognizing outstanding Ph.D. Research, Springer
- **2009:** Award for the most outstanding Ph.D. in Physics, University of Zaragoza
- **2005:**

  - Award for the most outstanding graduate in Physics, University of Zaragoza
  - Award for the best 30 degrees of the regional region of Aragón, by the Caja de Ahorros de la Inmaculada

Grants

- **2010:** Marie Crie Intra-European Fellowship (IEF) at the Cavendish Laboratory, University of Cambridge (UK), pending signature. Calling Institution: European Comission.
- **01/04/06 to 31/12/09:** Research Staff Training (F.P.U.) Ph.D. Grant at the Institute of Nanoscience of Aragón and the Institute of Material Sciences of Aragón. Calling institution: Spanish Ministry of Science and Innovation
- **01/01/2006 to 31/03/06:** Ph.D. Grant at the Institute of Nanoscience of Aragón and the Institute of Material Sciences of Aragón. Calling institution: Regional Government of Aragón
- **01/07/05 to 30/09/05:** Grant for the Introduction to Research (ultimate degree course) at the Department of Condensed Matter Physics of the University of Zaragoza. Calling institution: Spanish Scientific Research Council (C.S.I.C.)
- **10/09/04 to 31/07/05:** Collaboration in research Grant at the Department of Condensed Matter Physics of the University of Zaragoza. Calling institution: Spanish Ministry of Education and Science
- **01/07/04 to 30/09/04:** Grant for the Introduction to Research (penultimate degree course) at the Department of Condensed Matter Physics of the University of Zaragoza. Calling institution: Spanish Scientific Research Council (C.S.I.C.)

Stays in Scientific Institutions and Large Facilities

- **Advanced Light Source ALS, Lawrence Laboratory (Berkeley), United States of America.** Dates: 15/10/2009–13/11/2009. Local contact: Dr. Tolek

Tyliszczak (11.0.2. beamline). Topic: Scanning Transmission X-Ray Microscopy in cobalt nanoconstrictions

- **High Field Magnet Laboratory (Nijmegen), The Netherlands.** Dates: 26/10/08–01/11/08. Local contact: Dr. Uli Zeitler. Topic: Hall effect and magnetoresistance in high static magnetic fields (< 30 Tesla)
- **Imperial College London (London), United Kingdom.** Dates: 01/07/2008–1/09/2008. Scientist in charge: Prof. Russell Cowburn. Topic: Spatially-resolved MOKE measurements of Co-based nanodeposits grown by focused electron beam induced deposition
- **Instituto de Microelectrónica (Madrid), Spain.** Dates: 18/05/2006–20/05/2006. Scientist in charge: Prof. Ricardo García. Topic: Local oxidation by Atomic Force Microscopy of Fe_3O_4 thin films
- **Instituto de Catálisis y Petroleoquímica (Madrid), Spain.** Dates: 10/10/2005–25/10/2005. Scientist in charge: Prof. José Luis García Fierro. Topic: X-Ray Photoelectron Spectroscopy of iron oxides
- **Institute Max Von Laue et Paul Langevin, ILL (Grenoble), France.** Dates: 2005 (10 days). Local contact: Dr. Clemens Ritter (D11 beamline). Topic: Small angle neutron scattering on $La_{0.67}Ca_{0.33}Mn_{1-x}Ga_xO_3$.

Scientific Schools

- **SNOW School (Nancy), France.** Dates: 23/11/2008–25/11/2008. School Director: Prof. Stéphane Mangin (University of Nancy). Topic: Properties and techniques in magnetic single nano-objects.
- **New frontiers in Magnetism (Jaca), Spain.** Dates: 02/07/2007–07/07/2007. School director: Dr. L. M. García (University of Zaragoza), J. A. Blanco (University of Oviedo), J. Campo (CSIC). Topic: Present and future of magnetic materials.
- **II Panama Summer School (Toulouse), France.** Dates: 3/07/2006–14/07/2006. School Director: Prof. Christophe Vieu (LAAS–CNRS). Topic: Theory and practical training in emerging nanopatterning techniques.
- **II European School of Magnetism (Constanta), Romania.** Dates: 7/09/2005–16/09/2005. School Director: Dr. Stefania Pizzini (CNRS). Topic: New experimental approaches to magnetism
- **School of Neutron Scattering "Paolo Ricci" (Palau), Italy.** Dates: 21/09/2004–2/10/2004. School Director: Prof. Maria Antionetta Ricci (University of Rome). Topic: Small angle neutron scattering. Experiments and simulations

Technical Skills

- Focused ion beam (FIB)
- Scanning electron microscopy (SEM)
- Focused Electron/Ion beam Induced deposition (FEBID/FIBID)
- Photolithography techniques

- Magnetometry: Vibrating sample magnetometer (VSM), magneto-optical Kerr effect (MOKE)
- Four probe DC electrical transport measurements
- Thin film growth techniques: Magnetron sputtering, plasma enhanced chemical vapour deposition, electron-beam evaporation
- Ion milling
- Atomic Force Microscopy (AFM)
- Energy dispersive X-Ray spectroscopy (EDX), X-ray photoelectron spectroscopy (XPS)
- Intense static magnetic fields (9-30 Tesla)
- Neutron diffraction
- Micromagnetic simulations

Publications

(Articles: A, book chapters: BC)

A1. J. Orna, L. Morellón, P. A. Algarabel, J. M. De Teresa, A. Fernández-Pacheco, G. Simón, C. Magen, J. A. Pardo and M. R. Ibarra, "Fe_3O_4 epitaxial thin films and heterostructures: magnetotransport and magnetic properties", *CIIMTEC Ceramic Congress*(2010), to be published

A2. N. Marcano, S. Sangiao, M. Plaza, L. Pérez, A. Fernández Pacheco, R. Córdoba, M. C. Sánchez, L. Morellón, M. R. Ibarra and J. M. De Teresa, "Weak-antilocalization signatures in the magnetotransport properties of individual electrodeposited Bi Nanowires", *Applied Physics Letters* **96**, 082110 (2010)

A3. A. Fernández-Pacheco, J. M. De Teresa, P. A. Algarabel, J. Orna, L. Morellón, J. A. Pardo, and M. R. Ibarra, "Hall effect and magnetoresistance measurements in Fe_3O_4 thin films up to 30 Tesla", *Applied Physics Letters* **95**, 262108 (2009)

A4. A. Fernández-Pacheco, J. M. De Teresa, R. Córdoba, M. R. Ibarra, "Conduction regimes of Pt–C nanowires grown by Focused-Ion–Beam induced deposition: Metal-insulator transition", *Physical Review B* **79**, 174204 (2009)

A5. I. Guillamón, H. Suderow, A. Fernández-Pacheco, J. Sesé, R. Córdoba, J. M. De Teresa, M. R. Ibarra, and S. Vieira, "Direct observation of melting in a 2-D superconducting vortex lattice", *Nature Physics* **5**, 651 (2009)

A6. A. Fernández-Pacheco, J. M. De Teresa, R. Córdoba, M. R. Ibarra D. Petit, D. E. Read, L. O'Brien, E. R. Lewis, H. T. Zeng, and R. P. Cowburn, "Domain wall conduit behavior in cobalt nanowires grown by Focused-Electron-Beam Induced Deposition", *Applied Physics Letters* **94**, 192509 (2009)

A7. A. Fernández-Pacheco, J. M. De Teresa, R. Córdoba, M. R. Ibarra D. Petit, D. E. Read, L. O'Brien, E. R. Lewis, H. T. Zeng, and R. P. Cowburn, "Magnetization reversal in individual Co micro- and nano-wires grown by focused-electron-beam-induced deposition", *Nanotechnology* **20**, 475704 (2009)

A8. A. Fernández-Pacheco, J. M. De Teresa, R. Córdoba, M. R. Ibarra, "Magnetotransport properties of high-quality cobalt nanowires grown by focused-electron-beam-induced deposition", *Journal of Physics D: Applied Physics* **42**, 055005 (2009)

A9. A. Fernández-Pacheco, J. M. De Teresa, R. Córdoba, and M. R. Ibarra, "Tunneling and anisotropic-tunneling magnetoresistance in iron nanoconstrictions fabricated by focused-ion-beam", *MRS DD proceedings* **DD02-03**, 1181 (2009)

A10. J. M. De Teresa, R. Córdoba, A. Fernández-Pacheco, O. Montero, P. Strichovanec, and M. R. Ibarra, "Origin of the Difference in the Resistivity of As-Grown Focused-Ion and Focused-Electron-Beam-Induced Pt Nanodeposits", *Journal of Nanomaterials* **2009**, 936863 (2009)

A11. I. Guillamón, H. Suderow, S. Vieira, A. Fernández-Pacheco, J. Sesé, R. Córdoba, J. M. De Teresa, M. R. Ibarra, "Superconducting density of states at the border of an amorphous thin film grown by focused-ion-beam", *Journal of Physics: Conference Series* **152**, 052064 (2009)

A12. J. V. Oboňa, J. M. De Teresa, R. Córdoba, A. Fernández-Pacheco, and. M. R. Ibarra, "Creation of stable nanoconstrictions in metallic thin films via progressive narrowing by focused-ion-beam technique and in-situ control of resistance", *Microelectronic Engineering* **86**, 639 (2009)

A13. M. Plaza, L. Pérez, M. C. Sánchez, A. Fernández-Pacheco, and J. M. de. Teresa, "Electrodeposition of metastable Co-Cu solid solutions", *Solid State Communications* **149**, 2043 (2009)

A14. J. M. De Teresa, A. Fernández-Pacheco, R. Córdoba, J. Sesé, M. R. Ibarra, I. Guillamón, H. Suderow, S. Vieira, "Transport properties of superconducting amorphous W-based nanowires fabricated by focused-ion-beam-induced-deposition for applications in Nanotechnology", *MRS proceedings* **CC04-09**, 1880 (2009)

A15. A. Fernández-Pacheco, J. M. De Teresa, R. Córdoba, and M. R. Ibarra, "Exploring the conduction in atomic-sized metallic constrictions created by controlled ion etching", *Nanotechnology* **19**, 415302 (2008)

A16. A. Fernández-Pacheco, J. M. De Teresa, J. Orna, L. Morellón, P. A. Algarabel, J. A. Pardo, and M. R. Ibarra, "Universal Scaling of the Anomalous Hall Effect in Fe_3O_4 thin films", *Physical Review B* **77**, 100403(R) (2008)

A17. A. Fernández-Pacheco, J. M. De Teresa, J. Orna, L. Morellon, P. A. Algarabel, J. A. Pardo, M. R. Ibarra, C. Magen, and E. Snoeck, "Giant planar Hall effect in epitaxial Fe_3O_4 thin films and its temperature dependence", *Physical Review B* **78**, 212402 (2008)

A18. I. Guillamón, H. Suderow, S. Vieira, A. Fernández-Pacheco, J. Sesé, R. Córdoba, J. M. De Teresa, M. R. Ibarra, "Nanoscale superconducting properties of amorphous W-based deposits grown with a focused-ion-beam", *New Journal of Physics* **10**, 093005 (2008)

A19. J. M. De Teresa, A. Fernández-Pacheco, L. Morellón, J. Orna, J. A. Pardo, D. Serrate, P. A. Algarabel, and M. R. Ibarra, "Magnetotransport properties of Fe_3O_4 thin films for applications in Spin Electronics", *Microelectronic Engineering* **84**, 1660 (2007)

BC1. J. M. De Teresa, A. Fernández-Pacheco, R. Córdoba, and M. R. Ibarra, "Electrical transport properties of metallic nanowires and nanoconstrictions created with FIB/SEM dual beam". Book: "Nanofabrication using focused ion and electron beams: principles and applications". Editors: P. E. Russell, I. Utke, S. Moshkalev, *Oxford University Press*, 2010

BC2. M. Gabureac, I. Utke, J. M. De Teresa, and A. Fernandez-Pacheco, "Focused Ion and Electron Beam induced deposition of magnetic structures". Book: "Nanofabrication using focused ion and electron beams: principles and applications". Editors: P. E. Russell, I. Utke, S. Moshkalev, *Oxford University Press*, 2010

Contributions to Conferences and Workshops

In total, I have participated in over 40 international conferences or workshops. The following list is a selection of the most relevant oral conferences (O) and posters (P).

O1. (Invited) Title: Beyond Focused-ion Beam resolution by simultaneous electrical measurements. Congress: 2nd Kleindiek Users Meeting. Place: Tübingen (Germany). Year: 2010

O2. Title: Beyond FIB resolution by simultaneous electrical measurements. Congress: Microscience 2010-RMS. Place: London (UK). Year: 2010

O3. Title: Exploring the conduction in atomic-sized constrictions created by focused ion beams. Congress: MRS Spring meeting. Place: San Francisco (United States of America). Year: 2009

O4. Title: Nano-MOKE investigation on Co nanodeposits fabricated by focused electron beam deposition. Congress: 3rd Spanish workshop on nanolithography. Place: Madrid (Spain). Year: 2009

O5. Title: Magnetic properties of Cobalt nanodeposits created by Focused Electron Beam Deposition. Congress: 2nd Spanish workshop on nanolithography. Place: Barcelona (Spain). Year: 2008

O6. Title: Ion and electron beam induced deposits of platinum. Congress: EFUG users meeting. Place: Arcachon (France). Year: 2007

O7. Title: Electron Beam Induced Deposition of Cobalt in a Dual Beam system. Congress: 1st Spanish workshop on nanolithography. Place: Zaragoza (Spain). Year: 2007

O8. Title: Magnetotransport properties of epitaxial Fe_3O_4 thin films. Congress: THIOX. Place: Sant Feliu de Guixols (Spain). Year: 2007

P1. Title: Magnetic properties of Cobalt nanodeposits created by Focused Electron Beam Induced Deposition. Congress: MRS Spring meeting. Place: San Francisco (United States of America). Year: 2009

P2. Title: Exploring the conduction in atomic-sized constrictions created by focused ion beams. Congress: M-SNOW workshop. Place: Nancy (France). Year: 2009

P3. Title: High quality properties of cobalt nanowires grown by Focused-Electron-Beam-Induced Deposition: magnetotransport characterization. Congress: M-SNOW workshop. Place: Nancy (France). Year: 2009

P4. Title: Universal scaling of the Anomalous Hall effect in epitaxial Fe_3O_4 thin films. Congress: IEEE International Magnetics Conference. Place: Madrid (Spain). Year: 2008

P5. Title: Capabilities of the Nanolab equipment in Zaragoza (Spain). Congress: Third European FIB and DualBeam UserClub Meeting. Place: Eindhoven (The Netherlands). Year: 2007

P6. Title: Magnetotransport properties in epitaxial Fe_3O_4 thin films. Congress: Third International Conference on micro- and nano-engineering. Place: Istanbul (Turkey). Year: 2007

Participation in Research Projects

- Title: 3-Dimensional Spintronics (3SPIN).
  Funding Institution: European Research Council—Advanced Grant 2009 (ref. 247368).
  Duration: 01/03/2010–28/02/2015. Allocated money: 2.5 M€.
  Project Leader: Prof. Russell P. Cowburn
- Title: Nanotechnology based on hybrid graphene-magnetic/superconductor materials.
  Funding institution: Aragon Regional Government (PI046/09).
  Duration: 01/09/2009 – 31/10/2011. Allocated money: 70,000 €.
  Project Leader: Prof. José María De Teresa
- Title: Magnetic and magnetotransport properties of micro- and nano-structures.
  Funding institution: Spanish Ministry of Science and Innovation (MAT2008-06507-C02-02/NAN).
  Duration: 01/01/2009 – 31/12/2011. Allocated money: 121,000 €.
  Dr. José María De Teresa (ICMA) and Dr. Luis Morellón (INA)
- Title: Synthesis and characterization of magnetic materials for applications in magnetoelectronics and magnetic refrigeration.
  Funding institution: Spanish Inter-institutional Council of Science and Technology (MAT2005-05565-C02-02).
  Duration: 31/12/2005 – 30/12/2008. Allocated money: 119,000 €.
  Project Leaders: Dr. José María De Teresa (ICMA) and Dr. Luis Morellón (INA)

Teaching

- Master co-director of the graduate student Aleksandra Szkudlarek, title: "Micromagnetic study of cobalt nanostructures created by focused electron beam induced deposition", University of Zaragoza, Spain–Science and Technology University of Krakow, Poland (2009)
- Laboratory demonstrations, subject "Physics I", 1st course of Industrial Engineering, University of Zaragoza, 60 hours (2009)
- Laboratory demonstrations, subject "Physics I", 1st course of Architecture, University of Zaragoza, 30 hours (2009)
- Laboratory demonstrations, subject "Fundamentals of Physics", 1st course of Chemical Engineering, University of Zaragoza, 30 hours (2008)
- Participation in the Education day 2010-Imperial College London
- Supervisor of undergraduate students during Summer grants: Kishan Manani (Imperial College London, UK: 2010) and Luis Serrano (University of Zaragoza, Spain: 2008)
- Tutor of undergraduate students in the "Sixth week of introduction to research of the University of Zaragoza" (2006)

Other Merits

- Languages: English (fluent), French (basic), Spanish (native language)
- Courses for computing-1999: C, ADA, ASSEMBLER and SQL, University of Zaragoza, 200 hours
- Tutorial *"Focused Ion Beams for Analysis and Processing"*, MRS Fall meeting 2006, Boston (United Sates of America), 8 hours
- Basketball coach: 13 years of experience. Certificates—Spanish Basketball Federation, first and second level (2007 & 2010)